About the Author

Francesca Lazzeri, PhD, is an experienced scientist and machine learning practitioner with over a decade of both academic and industry experience. She currently leads an international team of cloud AI advocates and developers at Microsoft, managing a large portfolio of customers and building intelligent automated solutions on the cloud.

Francesca is an expert in big data technology innovations and the applications of machine learning–based solutions to real-world problems. Her work is unified by the twin goals of making better sense of microeconomic data and using those insights to optimize firm decision making. Her research has spanned the areas of machine learning, statistical modeling, and time series econometrics and forecasting as well as a range of industries—energy, oil and gas, retail, aerospace, healthcare, and professional services.

Before joining Microsoft, she was a research fellow at Harvard University in the Technology and Operations Management Unit. Francesca periodically teaches applied analytics and machine learning classes at universities and research institutions around the world. You can find her on Twitter @frlazzeri.

Machine Learning for Time Series Forecasting with Python®

Francesca Lazzeri, PhD

WILEY

About the Technical Editor

James York-Winegar holds a bachelor's degree in mathematics and physics and a master's degree in information and data science. He has worked in academia, healthcare, and technology consulting. James currently works with companies to enable machine learning workloads by enabling their data infrastructure, security, and metadata management. He also teaches machine learning courses at the University of California, Berkeley, focused on scaling up machine learning technology for big data.

Prior to leaving academia, James originally was focused on the cross section between experimental and theoretical physics and materials science. His research was focused on photo-structural transformations of non-oxide glasses or chalcogenide glasses. This introduced James to processing extremely large amounts of data and high-performance computing, where his work still leads him today.

James has had exposure to many industries through his consulting experience, including education, entertainment, commodities, finance, telecommunications, consumer packaged goods, startups, biotech, and technology. With this experience, he helps companies understand what is possible with their data and how to enable new capabilities or business opportunities. You can find his LinkedIn profile at linkedin.com/in/winegarj/.

Acknowledgments

In the past few years, I had the privilege to work with many data scientists, cloud advocates, developers, and professionals from Microsoft and Wiley: all these people inspired me and supported me through the creation and writing of this book. I am particularly grateful to the Cloud Advocacy team at Microsoft, for their trust and encouragement and for making my job so much easier and more enjoyable.

Thanks to Jim Minatel, associate publisher at John Wiley & Sons, who worked with me since the beginning of the publishing process and acted as a bridge between my work and the editorial staff. It was a pleasure to work with Pete Gaughan, content enablement manager; with David Clark, project editor, who managed the process that got us from outline to a finished manuscript; and with Saravanan Dakshinamurthy, content refinement specialist, who managed the last stages of this book development and made them go as smoothly as they did. I appreciate the support and guidance provided by the technical reviewer, James Winegar, and I hope we will work on other projects together in the future.

Last but not least, I will be forever grateful to my daughter, Nicole, who reminds me of the goodness in this world and inspires me to be the greatest version of myself; to my husband, Laurent, for his unconditional and endless support, inspiration, and encouragement; to my parents, Andrea and Anna Maria, and my brother, Marco, for always being there for me and believing in me through all walks of my life.

—Francesca Lazzeri

Contents at a Glance

Contents

Introduction

Time series data is an important source of information used for future decision making, strategy, and planning operations in different industries: from marketing and finance to education, healthcare, and robotics. In the past few decades, machine learning model-based forecasting has also become a very popular tool in the private and public sectors.

Currently, most of the resources and tutorials for machine learning model-based time series forecasting generally fall into two categories: code demonstration repo for certain specific forecasting scenarios, without conceptual details, and academic-style explanations of the theory behind forecasting and mathematical formula. Both of these approaches are very helpful for learning purposes, and I highly recommend using those resources if you are interested in understanding the math behind theoretical hypotheses.

This book fills that gap: in order to solve real business problems, it is essential to have a systematic and well-structured forecasting framework that data scientists can use as a guideline and apply to real-world data science scenarios. The purpose of this hands-on book is to walk you through the core steps of a practical model development framework for building, training, evaluating, and deploying your time series forecasting models.

The first part of the book (Chapters 1 and 2) is dedicated to the conceptual introduction of time series, where you can learn the essential aspects of time series representations, modeling, and forecasting.

In the second part (Chapters 3 through 6), we dive into autoregressive and automated methods for forecasting time series data, such as moving average, autoregressive integrated moving average, and automated machine learning for time series data. I then introduce neural networks for time series forecasting, focusing on concepts such as recurrent neural networks (RNNs) and the comparison of different RNN units. Finally, I guide you through the most important steps of model deployment and operationalization on Azure.

Along the way, I show at practice how these models can be applied to real-world data science scenarios by providing examples and using a variety of open-source Python packages and Azure. With these guidelines in mind, you should be ready to deal with time series data in your everyday work and select the right tools to analyze it.

What Does This Book Cover?

This book offers a comprehensive introduction to the core concepts, terminology, approaches, and applications of machine learning and deep learning for time series forecasting: understanding these principles leads to more flexible and successful time series applications.

In particular, the following chapters are included:

Chapter 1: Overview of Time Series Forecasting   This first chapter of the book is dedicated to the conceptual introduction of time series, where you can learn the essential aspects of time series representations, modeling, and forecasting, such as time series analysis and supervised learning for time series forecasting.

We will also look at different Python libraries for time series data and how libraries such as pandas, statsmodels, and scikit-learn can help you with data handling, time series modeling, and machine learning, respectively.

Finally, I will provide you with general advice for setting up your Python environment for time series forecasting.

Chapter 2: How to Design an End-to-End Time Series Forecasting Solution on the Cloud   The purpose of this second chapter is to provide an end-to-end systematic guide for time series forecasting from a practical and business perspective by introducing a time series forecasting template and a real-world data science scenario that we use throughout this book to showcase some of the time series concepts, steps, and techniques discussed.

Chapter 3: Time Series Data Preparation   In this chapter, I walk you through the most important steps to prepare your time series data for forecasting models. Good time series data preparation produces clean and well-curated data, which leads to more practical, accurate predictions.

Python is a very powerful programming language to handle data, offering an assorted suite of libraries for time series data and excellent support for time series analysis, such as SciPy, NumPy, Matplotlib, pandas, statsmodels, and scikit-learn.

You will also learn how to perform feature engineering on time series data, with two goals in mind: preparing the proper input data set that is compatible with the machine learning algorithm requirements and improving the performance of machine learning models.

Chapter 4: Introduction to Autoregressive and Automated Methods for Time Series Forecasting   In this chapter, you discover a suite of autoregressive methods for time series forecasting that you can test on your forecasting problems. The different sections in this chapter are structured to give you just enough information on each method to get started with a working code example and to show you where to look to get more information on the method.

We also look at automated machine learning for time series forecasting and how this method can help you with model selection and hyperparameter tuning tasks.

Chapter 5: Introduction to Neural Networks for Time Series Forecasting   In this chapter, I discuss some of the practical reasons data scientists may still want to think about deep learning when they build time series forecasting solutions. I then introduce recurrent neural networks and show how you can implement a few types of recurrent neural networks on your time series forecasting problems.

Chapter 6: Model Deployment for Time Series Forecasting   In this final chapter, I introduce Azure Machine Learning SDK for Python to build and run machine learning workflows. You will get an overview of some of the most important classes in the SDK and how you can use them to build, train, and deploy a machine learning model on Azure.

Through machine learning model deployment, companies can begin to take full advantage of the predictive and intelligent models they build and, therefore, transform themselves into actual AI-driven businesses.

Finally, I show how to build an end-to-end data pipeline architecture on Azure and provide deployment code that can be generalized for different time series forecasting solutions.

Reader Support for This Book

This book also features extensive sample code and tutorials using Python, along with its technical libraries, that readers can leverage to learn how to solve real-world time series problems.

Readers can access the sample code and notebooks at the following link:
`aka.ms/ML4TSFwithPython`

Companion Download Files

As you work through the examples in this book, the project files you need are all available for download from `aka.ms/ML4TSFwithPython`.

Each file contains sample notebooks and data that you can use to validate your knowledge, practice your technical skills, and build your own time series forecasting solutions.

How to Contact the Publisher

If you believe you've found a mistake in this book, please bring it to our attention. At John Wiley & Sons, we understand how important it is to provide our customers with accurate content, but even with our best efforts an error may occur.

In order to submit your possible errata, please email it to our customer service team at `wileysupport@wiley.com` with the subject line "Possible Book Errata Submission."

How to Contact the Author

We appreciate your input and questions about this book! You can find me on Twitter at @frlazzeri.

Overview of Time Series Forecasting

Time series is a type of data that measures how things change over time. In a time series data set, the *time* column does not represent a variable per se: it is actually a primary structure that you can use to order your data set. This primary temporal structure makes time series problems more challenging as data scientists need to apply specific data preprocessing and feature engineering techniques to handle time series data.

However, it also represents a source of additional knowledge that data scientists can use to their advantage: you will learn how to leverage this temporal information to extrapolate insights from your time series data, like trends and seasonality information, to make your time series easier to model and to use it for future strategy and planning operations in several industries. From finance to manufacturing and health care, time series forecasting has always played a major role in unlocking business insights with respect to time.

Following are some examples of problems that time series forecasting can help you solve:

- What are the expected sales volumes of thousands of food groups in different grocery stores next quarter?

- What are the resale values of vehicles after leasing them out for three years?

- What are passenger numbers for each major international airline route and for each class of passenger?

- What is the future electricity load in an energy supply chain infrastructure, so that suppliers can ensure efficiency and prevent energy waste and theft?

The plot in Figure 1.1 illustrates an example of time series forecasting applied to the energy load use case.

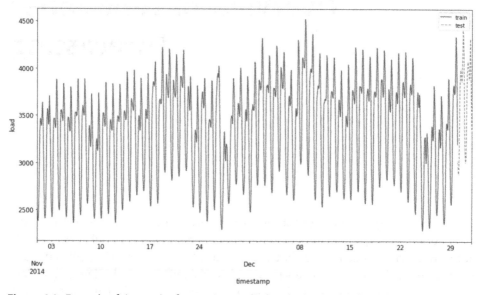

Figure 1.1: Example of time series forecasting applied to the energy load use case

This first chapter of the book is dedicated to the conceptual introduction— with some practical examples—of time series, where you can learn the essential aspects of time series representations, modeling, and forecasting.

Specifically, we will discuss the following:

- *Flavors of Machine Learning for Time Series Forecasting* – In this section, you will learn a few standard definitions of important concepts, such as time series, time series analysis, and time series forecasting, and discover why time series forecasting is a fundamental cross-industry research area.

- *Supervised Learning for Time Series Forecasting* – Why would you want to reframe a time series forecasting problem as a supervised learning problem? In this section you will learn how to reshape your forecasting scenario as a supervised learning problem and, as a consequence, get access to a large portfolio of linear and nonlinear machine learning algorithms.

- *Python for Time Series Forecasting* – In this section we will look at different Python libraries for time series data and how libraries such as pandas, statsmodels, and scikit-learn can help you with data handling, time series modeling, and machine learning, respectively.

- *Experimental Setup for Time Series Forecasting* – This section will provide you general advice for setting up your Python environment for time series forecasting.

Let's get started and learn some important elements that we must consider when describing and modeling a time series.

Flavors of Machine Learning for Time Series Forecasting

In this first section of Chapter 1, we will discover together why time series forecasting is a fundamental cross-industry research area. Moreover, you will learn a few important concepts to deal with time series data, perform time series analysis, and build your time series forecasting solutions.

One example of the use of time series forecasting solutions would be the simple extrapolation of a past trend in predicting next week hourly temperatures. Another example would be the development of a complex linear stochastic model for predicting the movement of short-term interest rates. Time-series models have been also used to forecast the demand for airline capacity, seasonal energy demand, and future online sales.

In time series forecasting, data scientists' assumption is that there is no causality that affects the variable we are trying to forecast. Instead, they analyze the historical values of a time series data set in order to understand and predict their future values. The method used to produce a time series forecasting model may involve the use of a simple deterministic model, such as a linear extrapolation, or the use of more complex deep learning approaches.

Due to their applicability to many real-life problems, such as fraud detection, spam email filtering, finance, and medical diagnosis, and their ability to produce actionable results, machine learning and deep learning algorithms have gained a lot of attention in recent years. Generally, deep learning methods have been developed and applied to univariate time series forecasting scenarios, where the time series consists of single observations recorded sequentially over equal time increments (Lazzeri 2019a).

For this reason, they have often performed worse than naïve and classical forecasting methods, such as exponential smoothing and autoregressive integrated moving average (ARIMA). This has led to a general misconception that deep learning models are inefficient in time series forecasting scenarios, and many data scientists wonder whether it's really necessary to add another class of methods, such as convolutional neural networks (CNNs) or recurrent neural networks (RNNs), to their time series toolkit (we will discuss this in more detail in Chapter 5, "Introduction to Neural Networks for Time Series Forecasting") (Lazzeri 2019a).

In time series, the chronological arrangement of data is captured in a specific column that is often denoted as *time stamp*, *date*, or simply *time*. As illustrated in Figure 1.2, a machine learning data set is usually a list of data points containing important information that are treated equally from a time perspective and are used as input to generate an output, which represents our predictions. On the contrary, a time structure is added to your time series data set, and all data points assume a specific value that is articulated by that temporal dimension.

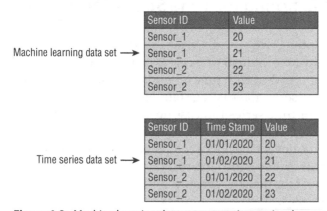

Figure 1.2: Machine learning data set versus time series data set

Now that you have a better understanding of time series data, it is also important to understand the difference between time series analysis and time series forecasting. These two domains are tightly related, but they serve different purposes: time series analysis is about identifying the intrinsic structure and extrapolating the hidden traits of your time series data in order to get helpful information from it (like trend or seasonal variation—these are all concepts that we will discuss later on in the chapter).

Data scientists usually leverage time series analysis for the following reasons:

- ▪ Acquire clear insights of the underlying structures of historical time series data.

- ▪ Increase the quality of the interpretation of time series features to better inform the problem domain.

- ▪ Preprocess and perform high-quality feature engineering to get a richer and deeper historical data set.

Time series analysis is used for many applications such as process and quality control, utility studies, and census analysis. It is usually considered the first step to analyze and prepare your time series data for the modeling step, which is properly called *time series forecasting*.

Time series forecasting involves taking machine learning models, training them on historical time series data, and consuming them to forecast future predictions. As illustrated in Figure 1.3, in time series forecasting that future output is unknown, and it is based on how the machine learning model is trained on the historical input data.

Sensor ID	Time Stamp	Values
Sensor_1	01/01/2020	60
Sensor_1	01/02/2020	64
Sensor_1	01/03/2020	66
Sensor_1	01/04/2020	65
Sensor_1	01/05/2020	67
Sensor_1	01/06/2020	68
Sensor_1	01/07/2020	70
Sensor_1	01/08/2020	69
Sensor_1	01/09/2020	72
Sensor_1	01/10/2020	?
Sensor_1	01/11/2020	?
Sensor_1	01/12/2020	?

Time series analysis on historical and/or current time stamps and values

Time series forecasting on future time stamps to generate future values

Figure 1.3: Difference between time series analysis historical input data and time series forecasting output data

Different historical and current phenomena may affect the values of your data in a time series, and these events are diagnosed as components of a time series. It is very important to recognize these different influences or components and decompose them in order to separate them from the data levels.

As illustrated in Figure 1.4, there are four main categories of components in time series analysis: *long-term movement or trend, seasonal short-term movements, cyclic short-term movements,* and *random or irregular fluctuations.*

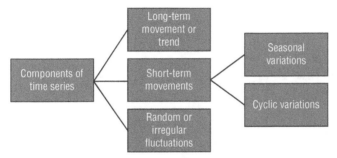

Figure 1.4: Components of time series

Let's have a closer look at these four components:

- *Long-term movement* or *trend* refers to the overall movement of time series values to increase or decrease during a prolonged time interval. It is common to observe trends changing direction throughout the course of your time series data set: they may increase, decrease, or remain stable at different moments. However, overall you will see one primary trend. Population counts, agricultural production, and items manufactured are just some examples of when trends may come into play.

- There are two different types of *short-term movements*:

  - *Seasonal variations* are periodic temporal fluctuations that show the same variation and usually recur over a period of less than a year. Seasonality is always of a fixed and known period. Most of the time, this variation will be present in a time series if the data is recorded hourly, daily, weekly, quarterly, or monthly. Different social conventions (such as holidays and festivities), weather seasons, and climatic conditions play an important role in seasonal variations, like for example the sale of umbrellas and raincoats in the rainy season and the sale of air conditioners in summer seasons.

  - *Cyclic variations*, on the other side, are recurrent patterns that exist when data exhibits rises and falls that are not of a fixed period. One complete period is a cycle, but a cycle will not have a specific predetermined length of time, even if the duration of these temporal fluctuations is usually longer than a year. A classic example of cyclic variation is a business cycle, which is the downward and upward movement of gross domestic product around its long-term growth trend: it usually can last several years, but the duration of the current business cycle is unknown in advance.

  As illustrated in Figure 1.5, cyclic variations and seasonal variations are part of the same short-term movements in time series forecasting, but they present differences that data scientists need to identify and leverage in order to build accurate forecasting models:

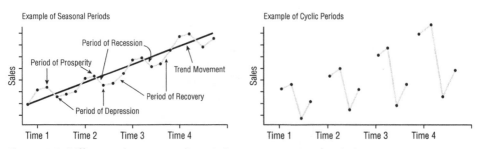

Figure 1.5: Differences between cyclic variations versus seasonal variations

■ *Random* or *irregular fluctuations* are the last element to cause variations in our time series data. These fluctuations are uncontrollable, unpredictable, and erratic, such as earthquakes, wars, flood, and any other natural disasters.

Data scientists often refer to the first three components (long-term movements, seasonal short-term movements, and cyclic short-term movements) as *signals* in time series data because they actually are deterministic indicators that can be derived from the data itself. On the other hand, the last component (random or irregular fluctuations) is an arbitrary variation of the values in your data that you cannot really predict, because each data point of these random fluctuations is independent of the other signals above, such as long-term and short-term movements. For this reason, data scientists often refer to it as *noise*, because it is triggered by latent variables difficult to observe, as illustrated in Figure 1.6.

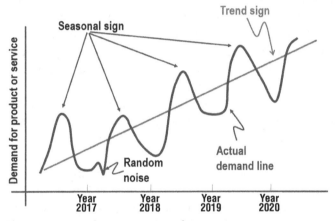

Figure 1.6: Actual representation of time series components

Data scientists need to carefully identify to what extent each component is present in the time series data to be able to build an accurate machine learning forecasting solution. In order to recognize and measure these four components, it is recommended to first perform a decomposition process to remove the component effects from the data. After these components are identified and measured, and eventually utilized to build additional features to improve the forecast accuracy, data scientists can leverage different methods to recompose and add back the components on forecasted results.

Understanding these four time series components and how to identify and remove them represents a strategic first step for building any time series forecasting solution because they can help with another important concept in time series that may help increase the predictive power of your machine learning algorithms: stationarity. *Stationarity* means that statistical parameters of a time series do not change over time. In other words, basic properties of the time series data distribution, like the mean and variance, remain constant over time. Therefore, stationary time series processes are easier to analyze and model because the basic assumption is that their properties are not dependent on time and will be the same in the future as they have been in the previous historical period of time. Classically, you should make your time series stationary.

There are two important forms of stationarity: *strong stationarity* and *weak stationarity*. A time series is defined as having a strong stationarity when all its statistical parameters do not change over time. A time series is defined as having a weak stationarity when its mean and auto-covariance functions do not change over time.

Alternatively, time series that exhibit changes in the values of their data, such as a trend or seasonality, are clearly not stationary, and as a consequence, they are more difficult to predict and model. For accurate and consistent forecasted results to be received, the nonstationary data needs to be transformed into stationary data. Another important reason for trying to render a time series stationary is to be able to obtain meaningful sample statistics such as means, variances, and correlations with other variables that can be used to get more insights and better understand your data and can be included as additional features in your time series data set.

However, there are cases where unknown nonlinear relationships cannot be determined by classical methods, such as autoregression, moving average, and autoregressive integrated moving average methods. This information can be very helpful when building machine learning models, and it can be used in feature engineering and feature selection processes. In reality, many economic time series are far from stationary when visualized in their original units of measurement, and even after seasonal adjustment they will typically still exhibit trends, cycles, and other nonstationary characteristics.

Time series forecasting involves developing and using a predictive model on data where there is an ordered relationship between observations. Before data scientists get started with building their forecasting solution, it is highly recommended to define the following forecasting aspects:

- *The inputs and outputs of your forecasting model* – For data scientists who are about to build a forecasting solution, it is critical to think about the data they have available to make the forecast and what they want to forecast about the future. Inputs are historical time series data provided to feed the model in order to make a forecast about future values. Outputs are the prediction results for a future time step. For example, the last seven days of energy consumption data collected by sensors in an electrical grid is considered input data, while the predicted values of energy consumption to forecast for the next day are defined as output data.

■ *Granularity level of your forecasting model* – Granularity in time series forecasting represents the lowest detailed level of values captured for each time stamp. Granularity is related to the frequency at which time series values are collected: usually, in Internet of Things (IoT) scenarios, data scientists need to handle time series data that has been collected by sensors every few seconds. IoT is typically defined as a group of devices that are connected to the Internet, all collecting, sharing, and storing data. Examples of IoT devices are temperature sensors in an air-conditioning unit and pressure sensors installed on a remote oil pump. Sometimes aggregating your time series data can represent an important step in building and optimizing your time series model: time aggregation is the combination of all data points for a single resource over a specified period (for example, daily, weekly, or monthly). With aggregation, the data points collected during each granularity period are aggregated into a single statistical value, such as the average or the sum of all the collected data points.

■ *Horizon of your forecasting model* – The horizon of your forecasting model is the length of time into the future for which forecasts are to be prepared. These generally vary from short-term forecasting horizons (less than three months) to long-term horizons (more than two years). Short-term forecasting is usually used in short-term objectives such as material requirement planning, scheduling, and budgeting; on the other hand, long-term forecasting is usually used to predict the long-term objectives covering more than five years, such as product diversification, sales, and advertising.

■ *The endogenous and exogenous features of your forecasting model* – Endogenous and *exogenous* are economic terms to describe internal and external factors, respectively, affecting business production, efficiency, growth, and profitability. *Endogenous features* are input variables that have values that are determined by other variables in the system, and the output variable depends on them. For example, if data scientists need to build a forecasting model to predict weekly gas prices, they can consider including major travel holidays as endogenous variables, as prices may go up because the cyclical demand is up.

On the other hand, *exogenous features* are input variables that are not influenced by other variables in the system and on which the output variable depends. Exogenous variables present some common characteristics (Glen 2014), such as these:

■ They are fixed when they enter the model.

■ They are taken as a given in the model.

- They influence endogenous variables in the model.
- They are not determined by the model.
- They are not explained by the model.

In the example above of predicting weekly gas prices, while the holiday travel schedule increases demand based on cyclical trends, the overall cost of gasoline could be affected by oil reserve prices, sociopolitical conflicts, or disasters such as oil tanker accidents.

- *The structured or unstructured features of your forecasting model* – Structured data comprises clearly defined data types whose pattern makes them easily searchable, while unstructured data comprises data that is usually not as easily searchable, including formats like audio, video, and social media postings. Structured data usually resides in relational databases, whose fields store length delineated data such as phone numbers, Social Security numbers, or ZIP codes. Even text strings of variable length like names are contained in records, making it a simple matter to search (Taylor 2018).

  Unstructured data has internal structure but is not structured via predefined data models or schema. It may be textual or non-textual, and human or machine generated. Typical human-generated unstructured data includes spreadsheets, presentations, email, and logs. Typical machine-generated unstructured data includes satellite imagery, weather data, landforms, and military movements.

  In a time series context, unstructured data doesn't present systematic time-dependent patterns, while structured data shows systematic time dependent patterns, such as trend and seasonality.

- *The univariate or multivariate nature of your forecasting model* – A univariate data is characterized by a single variable. It does not deal with causes or relationships. Its descriptive properties can be identified in some estimates such as central tendency (mean, mode, median), dispersion (range, variance, maximum, minimum, quartile, and standard deviation), and the frequency distributions. The univariate data analysis is known for its limitation in the determination of relationship between two or more variables, correlations, comparisons, causes, explanations, and contingency between variables. Generally, it does not supply further information on the dependent and independent variables and, as such, is insufficient in any analysis involving more than one variable.

  To obtain results from such multiple indicator problems, data scientists usually use multivariate data analysis. This multivariate approach will not only help consider several characteristics in a model but will also bring to light the effect of the external variables.

Time series forecasting can either be univariate or multivariate. The term *univariate time series* refers to one that consists of single observations recorded sequentially over equal time increments. Unlike other areas of statistics, the univariate time series model contains lag values of itself as independent variables (itl.nist.gov/div898/handbook/pmc/section4/pmc44.htm). These lag variables can play the role of independent variables as in multiple regression. The multivariate time series model is an extension of the univariate case and involves two or more input variables. It does not limit itself to its past information but also incorporates the past of other variables. Multivariate processes arise when several related time series are observed simultaneously over time instead of a single series being observed as in univariate case. Examples of the univariate time series are the ARIMA models that we will discuss in Chapter 4, "Introduction to Some Classical Methods for Time Series Forecasting." Considering this question with regard to inputs and outputs may add a further distinction. The number of variables may differ between the inputs and outputs; for example, the data may not be symmetrical. You may have multiple variables as input to the model and only be interested in predicting one of the variables as output. In this case, there is an assumption in the model that the multiple input variables aid and are required in predicting the single output variable.

▪ *Single-step or multi-step structure of your forecasting model* – Time series forecasting describes predicting the observation at the next time step. This is called a one-step forecast as only one time step is to be predicted. In contrast to the one-step forecast are the multiple-step or multi-step time series forecasting problems, where the goal is to predict a sequence of values in a time series. Many time series problems involve the task of predicting a sequence of values using only the values observed in the past (Cheng et al. 2006). Examples of this task include predicting the time series for crop yield, stock prices, traffic volume, and electrical power consumption. There are at least four commonly used strategies for making multi-step forecasts (Brownlee 2017):

  ▪ *Direct multi-step forecast*: The direct method requires creating a separate model for each forecast time stamp. For example, in the case of predicting energy consumption for the next two hours, we would need to develop a model for forecasting energy consumption on the first hour and a separate model for forecasting energy consumption on the second hour.

  ▪ *Recursive multi-step forecast*: Multi-step-ahead forecasting can be handled recursively, where a single time series model is created to forecast next time stamp, and the following forecasts are then computed using previous forecasts. For example, in the case of forecasting energy consumption for the next two hours, we would need to develop a one-step forecasting model. This model would then be used to predict next hour energy consumption, then this prediction would be used as input in order to predict the energy consumption in the second hour.

▪ *Direct-recursive hybrid multi-step forecast*: The direct and recursive strategies can be combined to offer the benefits of both methods (Brownlee 2017). For example, a distinct model can be built for each future time stamp, however each model may leverage the forecasts made by models at prior time steps as input values. In the case of predicting energy consumption for the next two hours, two models can be built, and the output from the first model is used as an input for the second model.

▪ *Multiple output forecast*: The multiple output strategy requires developing one model that is capable of predicting the entire forecast sequence. For example, in the case of predicting energy consumption for the next two hours, we would develop one model and apply it to predict the next two hours in one single computation (Brownlee 2017).

▪ *Contiguous or noncontiguous time series values of your forecasting model* – A time series that present a consistent temporal interval (for example, every five minutes, every two hours, or every quarter) between each other are defined as contiguous (Zuo et al. 2019). On the other hand, time series that are not uniform over time may be defined as noncontiguous: very often the reason behind noncontiguous timeseries may be missing or corrupt values. Before jumping to the methods of data imputation, it is important to understand the reason data goes missing. There are three most common reasons for this:

▪ *Missing at random*: Missing at random means that the propensity for a data point to be missing is not related to the missing data but it is related to some of the observed data.

▪ *Missing completely at random*: The fact that a certain value is missing has nothing to do with its hypothetical value and with the values of other variables.

▪ *Missing not at random*: Two possible reasons are that the missing value depends on the hypothetical value or the missing value is dependent on some other variable's value.

In the first two cases, it is safe to remove the data with missing values depending upon their occurrences, while in the third case removing observations with missing values can produce a bias in the model. There are different solutions for data imputation depending on the kind of problem you are trying to solve, and it is difficult to provide a general solution.

Moreover, since it has temporal property, only some of the statistical methodologies are appropriate for time series data.

I have identified some of the most commonly used methods and listed them as a structural guide in Figure 1.7.

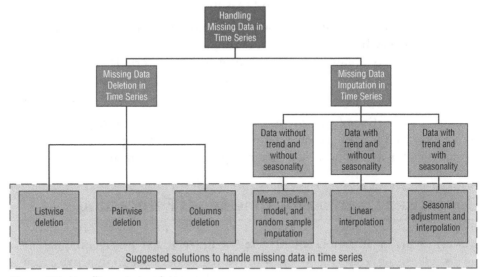

Figure 1.7: Handling missing data

As you can observe from the graph in Figure 1.7, listwise deletion removes all data for an observation that has one or more missing values. Particularly if the missing data is limited to a small number of observations, you may just opt to eliminate those cases from the analysis. However, in most cases it is disadvantageous to use listwise deletion. This is because the assumptions of the *missing completely at random* method are typically rare to support. As a result, listwise deletion methods produce biased parameters and estimates.

Pairwise deletion analyses all cases in which the variables of interest are present and thus maximizes all data available by an analysis basis. A strength to this technique is that it increases power in your analysis, but it has many disadvantages. It assumes that the missing data is missing completely at random. If you delete pairwise, then you'll end up with different numbers of observations contributing to different parts of your model, which can make interpretation difficult.

Deleting columns is another option, but it is always better to keep data than to discard it. Sometimes you can drop variables if the data is missing for more than 60 percent of the observations but only if that variable is insignificant. Having said that, imputation is always a preferred choice over dropping variables.

- Regarding time series specific methods, there are a few options:
    - *Linear interpolation*: This method works well for a time series with some trend but is not suitable for seasonal data.
    - *Seasonal adjustment and linear interpolation*: This method works well for data with both trend and seasonality.
    - *Mean, median, and mode*: Computing the overall mean, median, or mode is a very basic imputation method; it is the only tested function that takes no advantage of the time series characteristics or relationship between the variables. It is very fast but has clear disadvantages. One disadvantage is that mean imputation reduces variance in the data set.

In the next section of this chapter, we will discuss how to shape time series as a supervised learning problem and, as a consequence, get access to a large portfolio of linear and nonlinear machine learning algorithms.

Supervised Learning for Time Series Forecasting

Machine learning is a subset of artificial intelligence that uses techniques (such as deep learning) that enable machines to use experience to improve at tasks (`aka.ms/deeplearningVSmachinelearning`). The learning process is based on the following steps:

1. Feed data into an algorithm. (In this step you can provide additional information to the model, for example, by performing feature extraction.)

2. Use this data to train a model.

3. Test and deploy the model.

4. Consume the deployed model to do an automated predictive task. In other words, call and use the deployed model to receive the predictions returned by the model (`aka.ms/deeplearningVSmachinelearning`).

Machine learning is a way to achieve artificial intelligence. By using machine learning and deep learning techniques, data scientists can build computer systems and applications that do tasks that are commonly associated with human intelligence. These tasks include time series forecasting, image recognition, speech recognition, and language translation (`aka.ms/deeplearningVS-machinelearning`).

There are three main classes of machine learning: *supervised learning, unsupervised learning*, and *reinforcement learning*. In the following few paragraphs, we will have a closer look at each of these machine learning classes:

- *Supervised learning* is a type of machine learning system in which both input (which is the collection of values for the variables in your data set) and desired output (the predicted values for the target variable) are provided. Data is identified and labeled a priori to provide the algorithm a learning memory for future data handling. An example of a numerical label is the sale price associated with a used car (aka.ms/MLAlgorithmCS). The goal of supervised learning is to study many labeled examples like these and then to be able to make predictions about future data points, like, for example, assigning accurate sale prices to other used cars that have similar characteristics to the one used during the labeling process. It is called supervised learning because data scientists supervise the process of an algorithm learning from the training data set (www.aka.ms/MLAlgorithmCS): they know the correct answers and they iteratively share them with the algorithm during the learning process. There are several specific types of supervised learning. Two of the most common are *classification* and *regression*:

- *Classification*: Classification is a type of supervised learning used to identify what category new information belongs in. It can answer simple two-choice questions, like yes or no, true or false, for example:

  - Is this tweet positive?

  - Will this customer renew their service?

  - Which of two coupons draws more customers?

    Classification can be used also to predict between several categories and in this case is called multi-class classification. It answers complex questions with multiple possible answers, for example:

  - What is the mood of this tweet?

  - Which service will this customer choose?

  - Which of several promotions draws more customers?

- *Regression*: Regression is a type of supervised learning used to forecast the future by estimating the relationship between variables. Data scientists use it to achieve the following goals:

  - Estimate product demand

  - Predict sales figures

  - Analyze marketing returns

- *Unsupervised learning* is a type of machine learning system in which data points for the input have no labels associated with them. In this case, data is not labeled a priori so that the unsupervised learning algorithm itself can organize the data in and describe its structure. This can mean grouping it into clusters or finding different ways of looking at complex data structures (aka.ms/MLAlgorithmCS).

There are several types of unsupervised learning, such as cluster analysis, anomaly detection, and principal component analysis:

- *Cluster analysis*: Cluster analysis is a type of unsupervised learning used to separate similar data points into intuitive groups. Data scientists use it when they have to discover structures among their data, such as in the following examples:

  - Perform customer segmentation

  - Predict customer tastes

  - Determine market price

- *Anomaly detection*: Anomaly detection is a type of supervised learning used to identify and predict rare or unusual data points. Data scientists use it when they have to discover unusual occurrences, such as with these examples:

  - Catch abnormal equipment readings

  - Detect fraud

  - Predict risk

  The approach that anomaly detection takes is to simply learn what normal activity looks like (using a history of non-fraudulent transactions) and identify anything that is significantly different.

- *Principal component analysis*: Principal component analysis is a method for reducing the dimensionality of the feature space by representing it with fewer uncorrelated variables. Data scientists use it when they need to combine input features in order to drop the least important features while still retaining the most valuable information from the features in the data set.

  Principal component analysis is very helpful when data scientists need to answer questions such as the following:

  - How can we understand the relationships between each variable?

  - How can we look at all of the variables collected and focus on a few of them?

  - How can we avoid the danger of overfitting our model to our data?

- *Reinforcement learning* is a type of machine learning system in which the algorithm is trained to make a sequence of decisions. The algorithm learns to achieve a goal in an uncertain, potentially complex environment by employing a trial and error process to come up with a solution to the problem (aka.ms/MLAlgorithmCS).

Data scientists need to define the problem a priori, and the algorithm gets either rewards or penalties for the actions it performs. Its goal is to maximize the total reward. It's up to the model to figure out how to perform the task to maximize the reward, starting from totally random trials. Here are some examples of applications of reinforcement learning:

- Reinforcement learning for traffic signal control
- Reinforcement learning for optimizing chemical reactions
- Reinforcement learning for personalized news recommendations

When data scientists are choosing an algorithm, there are many different factors to take into consideration (`aka.ms/AlgorithmSelection`):

- *Evaluation criteria*: Evaluation criteria help data scientists to evaluate the performance of their solutions by using different metrics to monitor how well machine learning models represent data. They are an important step in the training pipeline to validate a model. There are different evaluation metrics for different machine learning approaches, such as accuracy, precision, recall, F-score, receiver operating characteristic (ROC), and area under the curve (AUC) for classification scenarios and mean absolute error (MAE), mean squared error (MSE), R-squared score, and adjusted R-squared for regression scenarios. MAE is a metric that can be used to measure forecast accuracy. As the name denotes, it is the mean of the absolute error: the absolute error is the absolute value of the difference between the forecasted value and the actual value, and it is scale-dependent: The fact that this metric is not scaled to the average demand can represent a limitation for data scientists who need to compare accuracy across time series with different scales. For time series forecasting scenarios, data scientists can also use the mean absolute percentage error (MAPE) to compare the fits of different forecasting and smoothing methods. This metric expresses accuracy as a percentage of MAE and allows data scientists to compare forecasts of different series in different scales.

- *Training time*: Training time is the amount of time needed to train a machine learning model. Training time is often closely tied to overall model accuracy. In addition, some algorithms are more sensitive to the number of data points than others. When time is limited, it can drive the choice of algorithm, especially when the data set is large.

- *Linearity*: Linearity is mathematical function that identifies a specific relationship between data points of a data set. This mathematical relationship means that data points can be graphically represented as a straight line. Linear algorithms tend to be algorithmically simple and fast to train. Different machine learning algorithms make use of linearity. Linear classification algorithms (such as logistic regression and support vector machines) assume that classes in a data set can be separated by a straight line. Linear regression algorithms assume that data trends follow a straight line.

▪ *Number of parameters*: Machine learning parameters are numbers (such as the number of error tolerance, the number of iterations, the number of options between variants of how the algorithm behaves) that data scientists usually need to manually select in order to improve an algorithm's performance (`aka.ms/AlgorithmSelection`). The training time and accuracy of the algorithm can sometimes be quite sensitive to getting just the right settings. Typically, algorithms with large numbers of parameters require the most trial and error to find a good combination. While this is a great way to make sure you've spanned the parameter space, the time required to train a model increases exponentially with the number of parameters. The upside is that having many parameters typically indicates that an algorithm has greater flexibility. It can often achieve very good accuracy, provided you can find the right combination of parameter settings (`aka.ms/AlgorithmSelection`).

▪ *Number of features*: Features are properties of a phenomenon based on which data scientists would like to predict results. A large number of features can overload some learning algorithms, making training time long. Data scientists can perform techniques such as feature selection and dimensionality reduction to reduce the number and the dimensionality of the features they have to work with. While both methods are used for reducing the number of features in a data set, there is an important difference:

  ▪ *Feature selection* is simply selecting and excluding given features without changing them.

  ▪ *Dimensionality reduction* transforms features into a lower dimension.

With these important machine learning concepts in mind, you can now learn how to reshape your forecasting scenario as a supervised learning problem and, as a consequence, get access to a large portfolio of linear and nonlinear machine learning algorithms

Time series data can be expressed as a supervised learning problem: data scientists usually transform their time series data sets into a supervised learning by exploiting previous time steps and using them as input and then leveraging the next time step as output of the model. Figure 1.8 shows the difference between an original time series data set and a data set transformed as a supervised learning.

We can summarize some observations from Figure 1.8 in the following way:

Time series data set

Sensor ID	Time Stamp	Value 1
Sensor_1	01/01/2020	236
Sensor_1	01/02/2020	133
Sensor_1	01/03/2020	148
Sensor_1	01/04/2020	152
Sensor_1	01/05/2020	241

Time series as supervised learning problem

Sensor ID	Time Stamp	Value x	Value y
Sensor_1	01/01/2020	NaN	236
Sensor_1	01/01/2020	236	133
Sensor_1	01/02/2020	133	148
Sensor_1	01/03/2020	148	152
Sensor_1	01/04/2020	152	241
Sensor_1	01/05/2020	241	Value to be predicted

Machine Learning
Prediction

Figure 1.8: Time series data set as supervised learning problem

- The value of Sensor_1 at prior time step (for example, 01/01/2020) becomes the input (Value x) in a supervised learning problem.

- The value of Sensor_1 at subsequent time step (for example, 01/02/2020) becomes the output (Value y) in a supervised learning problem.

- It is important to note that the temporal order between the Sensor_1 values needs to be maintained during the training of machine learning algorithms.

- By performing this transformation on our time series data, the resulting supervised learning data set will show an empty value (NaN) in the first row of Value x. This means that no prior Value x can be leveraged to predict the first value in the time series data set. We suggest removing this row because we cannot use it for our time series forecasting solution.

- Finally, the subsequent next value to predict for the last value in the sequence is unknown: this is the value that needs to be predicted by our machine learning model.

How can we turn any time series data set into a supervised learning problem? Data scientists usually exploit the values of prior time steps to predict the subsequent time step value by using a statistical method, called the *sliding window method*. Once the sliding window method is applied and a time series data set is converted, data scientists can use can leverage standard linear and nonlinear machine learning methods to model their time series data.

Previously and in Figure 1.8, I used examples of *univariate time series*: these are data sets where only a single variable is observed at each time, such as energy load at each hour. However, the sliding window method can be applied to a time series data set that includes more than one historical variable observed at each time step and when the goal is to predict more than one variable in the future: this type of time series data set is called *multivariate time series* (I will discuss this concept in more detail later in this book).

We can reframe this time series data set as a supervised learning problem with a window width of one. This means that we will use the prior time step values of Value 1 and Value 2. We will also have available the next time step value for Value 1. We will then predict the next time step value of Value 2. As illustrated in Figure 1.9, this will give us three input features and one output value to predict for each training pattern.

Multivariate time series data set

Sensor ID	Time Stamp	Value 1	Value 2
Sensor_1	01/01/2020	236	23
Sensor_1	01/02/2020	133	34
Sensor_1	01/03/2020	148	32
Sensor_1	01/04/2020	152	31
Sensor_1	01/05/2020	241	22

Multivariate time series as supervised learning problem

Sensor ID	Time Stamp	Value x	Value x2	Value x3	Value y
Sensor_1	01/01/2020	NaN	NaN	236	23
Sensor_1	01/01/2020	236	23	133	34
Sensor_1	01/02/2020	133	34	148	32
Sensor_1	01/03/2020	148	32	152	31
Sensor_1	01/04/2020	152	31	241	22
Sensor_1	01/05/2020	241	22	NaN	Value to be predicted

Machine Learning
Prediction

Figure 1.9: Multivariate time series as supervised learning problem

In the example of Figure 1.9, we were predicting two different output variables (Value 1 and Value 2), but very often data scientists need to predict multiple time steps ahead for one output variable. This is called *multi-step forecasting*. In multi-step forecasting, data scientists need to specify the number of time steps ahead to be forecasted, also called *forecast horizon* in time series. Multi-step forecasting usually presents two different formats:

- *One-step forecast*: When data scientists need to predict the next time step $(t + 1)$

- *Multi-step forecast*: When data scientists need to predict two or more (n) future time steps $(t + n)$

For example, demand forecasting models predict the quantity of an item that will be sold the following week and the following two weeks given the sales up until the current week. In the stock market, given the stock prices up until today one can predict the stock prices for the next 24 hours and 48 hours. Using a weather forecasting engine, one can predict the weather for the next day and for the entire week (Brownlee 2017).

The sliding window method can be applied also on a multi-step forecasting solution to transform it into a supervised learning problem. As illustrated in Figure 1.10, we can use the same univariate time series data set from Figure 1.8 as an example, and we can structure it as a two-step forecasting data set for supervised learning with a window width of one (Brownlee 2017).

Time series data set

Sensor ID	Time Stamp	Value 1
Sensor_1	01/01/2020	236
Sensor_1	01/02/2020	133
Sensor_1	01/03/2020	148
Sensor_1	01/04/2020	152
Sensor_1	01/05/2020	241

Time series as multi-step supervised learning

Sensor ID	Time Stamp	Value x	Value y	Value y2
Sensor_1	01/01/2020	NaN	236	133
Sensor_1	01/01/2020	236	133	148
Sensor_1	01/02/2020	133	148	152
Sensor_1	01/03/2020	148	152	241
Sensor_1	01/04/2020	152	241	NaN
Sensor_1	01/05/2020	241	Value to be predicted	Value to be predicted

Machine Learning Prediction   Machine Learning Prediction

Figure 1.10: Univariate time series as multi-step supervised learning

As illustrated in Figure 1.10, data scientists cannot use the first row (time stamp 01/01/2020) and the second to last row (time stamp 01/04/2020) of this sample data set to train their supervised model; hence we suggest removing it. Moreover, this new version of this supervised data set only has one variable Value x that data scientists can exploit to predict the last row (time stamp 01/05/2020) of both Value y and Value y^2.

In the next section you will learn about different Python libraries for time series data and how libraries such as pandas, statsmodels, and scikit-learn can help you with data handling, time series modeling, and machine learning, respectively.

Originally developed for financial time series such as daily stock market prices, the robust and flexible data structures in different Python libraries can be applied to time series data in any domain, including marketing, health care, engineering, and many others. With these tools you can easily organize, transform, analyze, and visualize your data at any level of granularity—examining details during specific time periods of interest and zooming out to explore variations on different time scales, such as monthly or annual aggregations, recurring patterns, and long-term trends.

Python for Time Series Forecasting

In this section, we will look at different Python libraries for time series data and how libraries such as pandas, statsmodels, and scikit-learn can help you with data handling, time series modeling, and machine learning, respectively. The Python ecosystem is the dominant platform for applied machine learning.

The primary rationale for adopting Python for time series forecasting is that it is a general-purpose programming language that you can use both for experimentation and in production. It is easy to learn and use, primarily because the language focuses on readability. Python is a dynamic language and very suited to interactive development and quick prototyping with the power to support the development of large applications.

Python is also widely used for machine learning and data science because of the excellent library support, and it has a few libraries for time series, such as NumPy, pandas, SciPy, scikit-learn, statsmodels, Matplotlib, datetime, Keras, and many more. Below we will have a closer look at the time series libraries in Python that we will use in this book:

- *SciPy*: SciPy is a Python-based ecosystem of open-source software for mathematics, science, and engineering. Some of the core packages are NumPy (a base n-dimensional array package), SciPy library (a fundamental library for scientific computing), Matplotlib (a comprehensive library for 2-D plotting), IPython (an enhanced interactive console), SymPy (a library for symbolic mathematics), and pandas (a library for data structure and analysis). Two SciPy libraries that provide a foundation for most others are NumPy and Matplotlib:

  - *NumPy* is the fundamental package for scientific computing with Python. It contains, among other things, the following:

    - A powerful n-dimensional array object

    - Sophisticated (broadcasting) functions

    - Tools for integrating C/C++ and Fortran code

    - Useful linear algebra, Fourier transform, and random number capabilities

      The most up-to-date NumPy documentation can be found at `numpy.org/devdocs/`. It includes a user guide, full reference documentation, a developer guide, meta information, and "NumPy Enhancement Proposals" (which include the NumPy Roadmap and detailed plans for major new features).

  - *Matplotlib*: Matplotlib is a Python plotting library that produces publication-quality figures in a variety of hardcopy formats and interactive environments across platforms. Matplotlib can be used in Python scripts, the Python and IPython shells, the Jupyter notebook, web application servers, and four graphical user interface toolkits.

Matplotlib is useful to generate plots, histograms, power spectra, bar charts, error charts, scatterplots, and so on with just a few lines of code. The most up-to-date Matplotlib documentation can be found in the Matplotlib user's guide (`matplotlib.org/3.1.1/users/index.html`).

Moreover, there are three higher-level SciPy libraries that provide the key features for time series forecasting in Python, they are pandas, statsmodels, and scikit-learn for data handling, time series modeling, and machine learning, respectively:

▪ *Pandas:* Pandas is an open-source, BSD-licensed library providing high performance, easy-to-use data structures, and data analysis tools for the Python programming language. Python has long been great for data munging and preparation, but less so for data analysis and modeling. Pandas helps fill this gap, enabling you to carry out your entire data analysis workflow in Python without having to switch to a more domain-specific language like R.

The most up-to-date pandas documentation can be found in the pandas user's guide (`pandas.pydata.org/pandas-docs/stable/`).

Pandas is a NumFOCUS sponsored project. This will help ensure the success of development of pandas as a world-class open-source project.

Pandas does not implement significant modeling functionality outside of linear and panel regression; for this, look to statsmodels and scikit-learn below.

▪ *Statsmodels:* Statsmodels is a Python module that provides classes and functions for the estimation of many different statistical models as well as for conducting statistical tests and statistical data exploration. An extensive list of result statistics is available for each estimator. The results are tested against existing statistical packages to ensure that they are correct. The package is released under the open-source Modified BSD (3-clause) license.

The most up-to-date statsmodels documentation can be found in the statsmodels user's guide (`statsmodels.org/stable/index.html`).

▪ *Scikit-learn* - Scikit-learn is a simple and efficient tool for data mining and data analysis. In particular, this library implements a range of machine learning, pre-processing, cross-validation, and visualization algorithms using a unified interface. It is built on NumPy, SciPy, and Matplotlib and is released under the open-source Modified BSD (3-clause) license.

Scikit-learn is focused on machine learning data modeling. It is not concerned with the loading, handling, manipulating, and visualizing of data. For this reason, data scientists usually combine using scikit-learn with other libraries, such as NumPy, pandas, and Matplotlib, for data handling, pre-processing, and visualization.

The most up-to-date scikit-learn documentation can be found in the scikit-learn user's guide (`scikit-learn.org/stable/index.html`).

In this book, we will also use Keras for time series forecasting:

- *Keras:* Keras is a high-level neural networks API, written in Python and capable of running on top of TensorFlow, CNTK, and Theano. Data scientists usually use Keras if they need a deep learning library that does the following:

  - Allows for easy and fast prototyping (through user friendliness, modularity, and extensibility)

  - Supports both convolutional networks and recurrent networks, as well as combinations of the two

  - Runs seamlessly on central processing unit (CPU) and graphics processing unit (GPU)

  The most up-to-date Keras documentation can be found in the Keras user's guide (`keras.io`).

Now that you have a better understanding of the different Python packages that we will use in this book to build our end-to-end forecasting solution, we can move to the next and last section of this chapter, which will provide you with general advice for setting up your Python environment for time series forecasting.

Experimental Setup for Time Series Forecasting

In this section, you will learn how to get started with Python in Visual Studio Code and how to set up your Python development environment. Specifically, this tutorial requires the following:

- *Visual Studio Code:* Visual Studio Code (VS Code) is a lightweight but powerful source code editor that runs on your desktop and is available for Windows, macOS, and Linux. It comes with built-in support for JavaScript, TypeScript, and Node.js and has a rich ecosystem of extensions for other languages (such as C++, C#, Java, Python, PHP, Go) and runtimes (such as .NET and Unity).

- *Visual Studio Code Python extension:* Visual Studio Code Python extension is a Visual Studio Code extension with rich support for the Python language (for all actively supported versions of the language: 2.7, ≥ 3.5), including features such as IntelliSense, linting, debugging, code navigation, code formatting, Jupyter notebook support, refactoring, variable explorer, and test explorer.

- *Python 3:* Python 3.0 was originally released in 2008 and is the latest major version of the language, with the latest version of the language, Python 3.8, being released in October 2019. In most of our examples in this book, we will use Python version 3.8.

- It is important to note that Python 3.x is incompatible with the 2.x line of releases. The language is mostly the same, but many details, especially how built-in objects like dictionaries and strings work, have changed considerably, and a lot of deprecated features were finally removed. Here are some Python 3.0 resources:

  - Python documentation (`python.org/doc/`)
  - Latest Python updates (`aka.ms/PythonMS`)

If you have not already done so, install VS Code. Next, install the Python extension for VS Code from the Visual Studio Marketplace. For additional details on installing extensions, see Extension Marketplace. The Python extension is named Python and published by Microsoft.

Along with the Python extension, you need to install a Python interpreter, following the instructions below:

- If you are using Windows:

  - Install Python from `python.org`. You can typically use the Download Python button that appears first on the page to download the latest version.

  - *Note:* If you don't have admin access, an additional option for installing Python on Windows is to use the Microsoft Store. The Microsoft Store provides installs of Python 3.7 and Python 3.8. Be aware that you might have compatibility issues with some packages using this method.

  - For additional information about Python on Windows, see *Using Python on Windows* at `python.org`.

- If you are using macOS:

  - The system installation of Python on macOS is not supported. Instead, an installation through Homebrew is recommended. To install Python using Homebrew on macOS use `brew install python3` at the Terminal prompt.

- ▪ *Note:* On macOS, make sure the location of your VS Code installation is included in your PATH environment variable. See these setup instructions for more information.

- ▪ If you are using Linux:

  - ▪ The built-in Python 3 installation on Linux works well, but to install other Python packages you must install pip with `get-pip.py`.

To verify that you have installed Python successfully on your machine, run one of the following commands (depending on your operating system):

- ▪ *Linux/macOS*: Open a Terminal Window and type the following command:

```
python3 --version
```

- ▪ *Windows*: Open a command prompt and run the following command:

```
py -3 --version
```

If the installation was successful, the output window should show the version of Python that you installed.

Conclusion

In this chapter, I walked you through the core concepts and steps to prepare your time series data for forecasting models. Through some practical examples of time series, we discussed some essential aspects of time series representations, modeling, and forecasting.

Specifically, we discussed the following topics:

- ▪ *Flavors of Machine Learning for Time Series Forecasting*: In this section you learned a few standard definitions of important concepts, such as time series, time series analysis, and time series forecasting. You also discovered why time series forecasting is a fundamental cross-industry research area.

- ▪ *Supervised Learning for Time Series Forecasting*: In this section you learned how to reshape your forecasting scenario as a supervised learning problem and, as a consequence, get access to a large portfolio of linear and nonlinear machine learning algorithms.

■ *Python for Time Series Forecasting*: In this section we looked at different Python libraries for time series data such as pandas, statsmodels, and scikit-learn.

■ *Experimental Setup for Time Series Forecasting*: This section provided you with a general guide for setting up your Python environment for time series forecasting.

In the next chapter, we will discuss some practical concepts such as the time series forecast framework and its applications. Moreover, you will learn about some of the caveats that data scientists working on forecasting projects may face. Finally, I will introduce a use case and some key techniques for building machine learning forecasting solutions successfully.

How to Design an End-to-End Time Series Forecasting Solution on the Cloud

As we discussed in Chapter 1, "Overview of Time Series Forecasting," time series forecasting is a method for the prediction of events through a sequence of time, by studying the behavior and performance of past phenomena and assuming that future occurrences will hold similar to historical trends and behaviors.

Nowadays time series forecasting is performed in a variety of applications, including weather forecasting, earthquake prediction, astronomy, finance, and control engineering. In many modern and real-world applications, time series forecasting uses computer technologies, including cloud, artificial intelligence, and machine learning, to build and deploy end-to-end forecasting solutions.

To solve real business problems in the industry, it is essential to have a systematic and well-structured template that data scientists can use as a guideline and can apply it to solve real-world business scenarios. The purpose of this second chapter is to provide an end-to-end systematic guide for time series forecasting from a practical perspective by introducing the following concepts:

- *Time Series Forecasting Template*: A time series forecast template is a set of tasks that leads from defining a business problem through to the outcome of having a time series forecast model deployed and ready to be consumed externally or across your company.

In the first section, I will introduce some key techniques for building machine learning forecasting solutions successfully. Specifically, we will take a closer look at the following steps:

▪ *Business understanding and performance metrics definition*: The business understanding performance metrics definition step outlines the business aspect we need to understand and consider prior to making an investment decision. It explains how to qualify the business problem at hand to ensure that predictive analytics and machine learning are indeed effective and applicable.

▪ *Data ingestion:* Data ingestion is the process of collecting and importing data to be cleaned and analyzed or stored in a database. Businesses today gather large volumes of data, both structured and unstructured, in an effort to use that data to discover real-time or near real-time insights.

▪ *Data exploration and understanding:* Once the raw data has been ingested and securely stored, it is ready to be explored. The data exploration and understanding phase is about taking your raw data and converting it into a form that can be used for data cleaning and feature engineering.

▪ *Data pre-processing and feature engineering:* Data pre-processing and feature engineering is the step where data scientists clean data sets from outliers and missing data and create additional features with the raw data to feed their machine learning models. In machine learning, a feature is a quantifiable variable of the phenomenon you are trying to analyze. For certain types of data, the number of features can be very large compared to the number of data points.

▪ *Modeling building and selection:* The modeling phase is where the conversion of the data into a model takes place. In the core of this process there are advanced algorithms that scan the historical data (training data), extract patterns, and build a model. That model can be later used to predict on new data that has not been used to build the model.

▪ *Model deployment:* Deployment is the method by which you integrate a machine learning model into an existing production environment in order to start using it to make practical business decisions based on data. It is one of the last stages in the machine learning life cycle.

▪ *Forecasting solution acceptance:* In the last stage of the time series forecasting solution deployment, data scientists need to confirm and validate that the pipeline, the model, and their deployment in a production environment satisfy their success criteria.

▪ *Use case: demand forecasting:* At the end of this chapter, I will introduce a real-world data science scenario that will be used throughout this book to showcase some of the time series concepts, steps, and techniques discussed. I believe everyone must learn to smartly work with huge amounts of data, hence large data sets are included, open and free to access.

Let's now get started and discover together how you can apply this time series forecasting template to your time series data and solutions.

Time Series Forecasting Template

This template is an agile and iterative framework to deliver a time series forecasting solution efficiently. It contains a distillation of the best practices and structures that facilitate the successful implementation of time series forecasting initiatives. The goal is to help companies fully realize the benefits of their data and build end-to-end forecasting solutions on the cloud.

To handle the increasing variety and complexity of forecasting problems, many machine learning and deep learning forecasting techniques have been developed in recent years. As we will discuss in Chapter 4, "Introduction to Some Classical Methods for Time Series Forecasting," and Chapter 5, "Introduction to Neural Networks for Time Series Forecasting," each of these forecasting techniques has its special application, and care must be taken to select the correct technique for a particular application. The better you understand the portfolio of algorithms that you can apply to your forecasting scenario, the more likely it is that your forecasting efforts will be successful.

The choice of a machine learning algorithm depends on many factors—the business question you are trying to answer, the relevance and availability of historical data, the accuracy and success metric you need to achieve, the horizon and how much time your team has to build the forecasting solution. These constraints must be weighed continuously, and on different levels (Lazzeri 2019b).

Figure 2.1 illustrates a time series forecasting template: the purpose here is to present an overview of this field by discussing the way a data scientist or a company should approach a forecasting problem, introducing a few approaches, describing the methods available, and explaining how to match each step to different aspects of a forecasting problem.

As you can observe in Figure 2.1, our template consists of different steps:

1. Business understanding and performance metrics definition

2. Data ingestion

3. Data exploration and understanding

4. Data pre-processing and feature engineering

5. Model building and selection

6. Model deployment

7. Forecasting solution acceptance

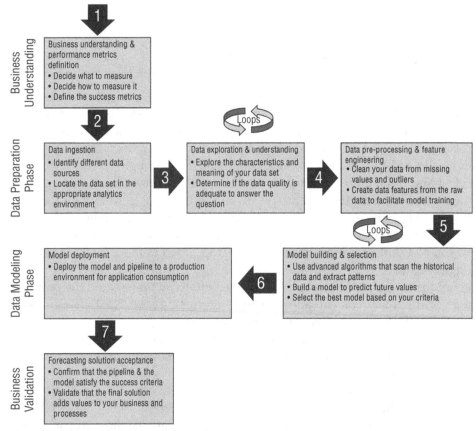

Figure 2.1: Time series forecasting template

There are some necessary iterative loops (on steps 3 and 5) that you should keep in mind within this process:

■ *Data exploration and understanding* – Data exploration and understanding is the moment in which data scientists will use several different statistical techniques and visualizations to explore the data and get a better understanding of its nature. It is possible that at this stage, data scientists will have additional questions and will need to request and ingest additional or different data and reevaluate the performance metrics defined at the beginning of this process.

- *Model building and selection* – During the model building and selection phase, it is possible that data scientists will have additional ideas in terms of new features to build and include in the models, new parameters to try, or even new algorithms to apply.

In the next few sections of this chapter, we will take a closer look at these different steps and introduce you to the use cases that we will be using throughout the book.

Business Understanding and Performance Metrics

The business understanding and performance metrics definition step outlines the business aspect we need to understand and consider prior to making an investment decision. It explains how to qualify the business problem at hand to ensure that predictive analytics and machine learning are indeed effective and applicable. For most organizations, lack of data is not a problem. In fact, it is the opposite: there is often too much information available to make a clear decision on the future.

With so much data to sort through, organizations need a well-defined strategy to clarify the following business aspects:

- How can forecasting help organizations transform business, better manage costs, and drive greater operational excellence?

- Do organizations have a well-defined and clearly articulated purpose and vision for what they are looking to accomplish in the future?

- How can organizations get support of C-level executives and stakeholders to take that forecasting approach and data-driven vision and drive them through the different parts of a business?

Companies need to have a clear understanding of their business and decision-making process to support the forecasting solution. The goal of this first step is to specify the key performance metrics, data, and features that will serve as the model targets and whose related metrics will be used to determine the success of the project, and also identify the relevant data sources that the business has access to or needs to obtain.

With the right mindset, what was once an overwhelming volume of disparate information becomes a simple, clear decision point. Organizations must begin with the right questions. Questions should be measurable, clear and concise, and directly correlated to their core business. In this stage, it is important to design questions to either qualify or disqualify potential solutions to a specific business problem or opportunity. For example, start with a clearly defined problem: a retail company is experiencing rising costs and is no longer able to offer competitive prices to its customers. One of many questions to solve this business problem might include, "Can the company reduce its operations without compromising quality?"

There are two main tasks that organizations need to address to answer those type of questions (Lazzeri 2019b):

- *Define business goals*: Companies need to work with business experts and other stakeholders to understand and identify the business problems.

- *Formulate right questions*: Companies need to formulate tangible questions that define the business goals that the forecasting teams can target.

To successfully translate this vision and business goal into actionable results, the next step is to establish clear performance metrics. In this second step, organizations need to focus on these two analytical aspects (Lazzeri 2019b):

- What is the best forecasting approach to tackle that business problem and draw accurate conclusions?

- How can that vision be translated into actionable results able to improve a business? This step breaks down into three sub-steps:

  - Decide what to measure.
  - Decide how to measure it.
  - Define the success metrics.

Let's take a closer look at each of these sub-steps:

- *Decide what to measure* – Let's take *predictive maintenance*, a technique used to predict when an in-service machine will fail, allowing for its maintenance to be planned well in advance. As it turns out, this is a very broad area with a variety of end goals, such as predicting root causes of failure, identifying which parts will need replacement and when, providing maintenance recommendations after the failure happens, and so on (Lazzeri 2019b).

  Many companies are attempting predictive maintenance and have piles of data available from all sorts of sensors and systems. But, too often, customers do not have enough data about their failure history, and that makes it is very difficult to do predictive maintenance—after all, models need to be trained on such failure history data in order to predict future failure incidents.

So, while it's important to lay out the vision, purpose, and scope of any analytics projects, it is critical that you start off by gathering the right data. Once you have a clear understanding of what you will measure, you will need to decide how to measure it and what the best analytical technique is in order to achieve this goal. Next we will discuss how you can make these decisions and pick the best machine learning approach to solve your business problem (Lazzeri 2019b).

▪ *Decide how to measure it* – Thinking about how organizations measure their data is just as important, especially before the data collection and ingestion phase. Key questions to ask for this sub-step are as follows:

- ▪ What is the time frame?

- ▪ What is the unit of measure?

- ▪ What factors should be included?

A central objective of this step is to identify the key business variables that the forecasting analysis needs to predict. We refer to these variables as the model targets, and we use the metrics associated with them to determine the success of the project. Two examples of such targets are sales forecasts or the probability of an order being fraudulent.

Once you have decided on the best machine learning approach to solve your business problem and identified the key business variables that the forecasting analysis needs to predict, you need to define a few success metrics that will help you define your project success.

▪ *Define the success metrics* – After the key business variables identification, it is important to translate a business problem into a data science question and define the metrics that will define a project success. At this point it is also important to revisit the project goals by asking and refining sharp questions that are relevant, specific, and unambiguous. With this data, companies can offer customer promotions to reduce churn. An example that comes up frequently: an organization may be interested in making a business justification for a given use case in which a cloud-based solution and machine learning are important components (Lazzeri 2019b).

Unlike an on-premises solution, in the case of a cloud-based solution, the upfront cost component is minimal and most of the cost elements are associated with actual usage. An organization should have a good understanding of the business value of operating a forecasting solution (short or long term). In fact, it is important to realize the business value of each forecast operation. In an energy demand forecasting example, accurately forecasting power load for the next 24 hours can prevent overproduction or can help prevent overloads on the grid, and this can be quantified in terms of financial savings on a daily basis.

Here is a basic formula for calculating the financial benefit of demand forecast solution:

$$\left(\frac{Cost\ of\ Storage + Cost\ of\ Data\ Egress + Cost\ of\ Forecast\ Transaction}{Number\ of\ Forecast\ Transactions} \right)$$
$$= \frac{Financial\ Value\ of\ Forecast}{Number\ of\ Forecast\ Transactions}$$

Nowadays, organizations create, acquire and store a combination of structured, semistructured, and unstructured data that can be mined for insights and used in machine learning applications to improve operations. Hence, it is crucial for companies to learn how to successfully manage the process of obtaining and importing data for immediate use or storage in a database.

As discussed in the energy example above, companies today rely on data to make all kinds of decisions—predict trends, forecast the market, plan for future needs, and understand their customers. However, how do you get all your company's data in one place so you can make the right decisions?

In the next section we will discuss a few techniques and best practices for companies to support data ingestion processes and move their data from multiple different sources into one place.

Data Ingestion

Businesses today gather large volumes of data, both structured and unstructured, in an effort to use that data to discover real-time or near real-time insights that inform decision making and support digital transformation. Data ingestion is the process of obtaining and importing data for immediate use or storage in a database.

Data can be ingested using three different methods: batch, real-time, and streaming:

- *Batch time series data processing* – A common big data scenario is batch processing of data at rest. In this scenario, the source data is loaded into data storage, either by the source application itself or by an orchestration workflow. The data is then processed in-place by a parallelized job, which can also be initiated by the orchestration workflow, as illustrated in Figure 2.2.

  The processing may include multiple iterative steps before the transformed results are loaded into an analytical data store, which can be queried by analytics and reporting components. To learn more about how to choose a batch processing technology, you can read the article on Azure Time Series Insights at `aka.ms/TimeSeriesInsights`.

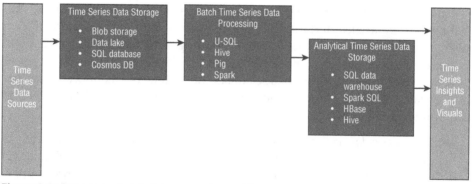

Figure 2.2: Time series batch data processing architecture

■ *Real-time data processing* – Real-time processing deals with streams of data that are captured in real time and processed with minimal latency to generate real-time (or near-real-time) reports or automated responses. For example, a real-time traffic monitoring solution might use sensor data to detect high traffic volumes. This data could be used to dynamically update a map to show congestion or automatically initiate high-occupancy lanes or other traffic management systems. To learn more about technologies that are recommended choices for real-time processing solutions, you can read the article at `aka.ms/RealTimeProcessing`.

■ *Streaming data* – With streaming data, data is immediately processed off of incoming data flows (`aka.ms/RealTimeProcessing`). After capturing real-time messages, the solution must process them by filtering, aggregating, and otherwise preparing the data for analysis, as illustrated in Figure 2.3.

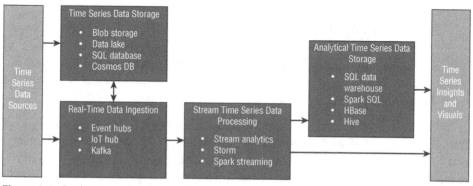

Figure 2.3: Real-time and streaming data processing architecture

An effective data ingestion process begins by prioritizing data sources, validating individual files, and routing data items to the correct destination. Moreover, a data ingestion pipeline moves streaming data and batched data from preexisting databases and data warehouses to a data lake.

In the data collection phase, data scientists and architects need to work together and usually evaluate two different types of tools: time series data collection tools and storage tools.

- *Time series data collection tools* – Time series data collection tools are the ones that will help you in the extraction and organization of raw data. An example is Stream Analytics, which is a real-time analytics and complex event-processing engine that is designed to analyze and process high volumes of fast streaming data from multiple sources simultaneously (`aka.ms/AzureStreamAnalytics`). Patterns and relationships can be identified in information extracted from a number of input sources including devices, sensors, clickstreams, social media feeds, and applications. These patterns can be used to trigger actions and initiate workflows such as creating alerts, feeding information to a reporting tool, or storing transformed data for later use.

  Another example of a data collection tool is Data Factory, a managed cloud service that is built for complex hybrid extract-transform-load (ETL), extract-load-transform (ELT), and data integration projects. It allows you to create data-driven workflows for orchestrating data movement and transforming data at scale (`aka.ms/AzureStreamAnalytics`).

- *Data storage tools* – Storage tools are databases that allow you to store your data. These tools store data in either structured or unstructured form and can aggregate information from several platforms in an integrated manner, such as Cosmos DB: today's applications are required to be highly responsive and always online (`aka.ms/MicrosoftCosmosDB`). To achieve low latency and high availability, instances of these applications need to be deployed in datacenters that are close to their users. Cosmos DB is a globally distributed, multi-model database service, that enables you to scale throughput and storage elastically and independently across many regions worldwide (`www.aka.ms/MicrosoftCosmosDB`).

In the case of demand forecasting, load data needs to be constantly and frequently predicted, and we must ensure that the raw data is flowing by means of a solid and reliable data ingestion process. The ingestion process must guarantee that the raw data is available for the forecasting process at the required time. That implies that the data ingestion frequency should be greater than the forecasting frequency.

If our demand forecasting solution would generate a new forecast at 8:00 a.m. on a daily basis, then we need to ensure that all the data that has been collected during the last 24 hours has been fully ingested till that point and has to include the last hour of data.

Now that you have a better understanding of data ingestion, we are ready for performing data exploration. In the next section, we will take a closer look at different techniques to explore data in order to bring important aspects of that data into focus for further time series analysis.

Data Exploration and Understanding

Data exploration is the first step in data analysis and typically involves summarizing the main characteristics of a data set, including its size, accuracy, initial patterns in the data, and other attributes. The raw data source that is required for performing reliable and accurate forecasting must meet some basic data quality criteria. Although advanced statistical methods can be used to compensate for some possible data quality issue, we still need to ensure that we are crossing some base data quality threshold when ingesting new data (Lazzeri 2019b).

Here are a few considerations concerning raw data quality:

- *Missing value* – This refers to the situation when specific measurement was not collected. The basic requirement here is that the missing value rate should not be greater than 10 percent for any given time period. In a case in which a single value is missing, it should be indicated by using a predefined value (for example, '9999') and not '0,' which could be a valid measurement.

- *Measurement accuracy* – The actual value of consumption or temperature should be accurately recorded. Inaccurate measurements will produce inaccurate forecasts. Typically, the measurement error should be lower than 1 percent relative to the true value.

- *Time of measurement* – It is required that the actual time stamp of the data collected will not deviate by more than 10 seconds relative to the true time of the actual measurement.

- *Synchronization* – When multiple data sources are being used (for example, consumption and temperature), we must ensure that there are no time synchronization issues between them. This means that the time difference between the collected time stamp from any two independent data sources should not exceed more than 10 seconds.

- *Latency* – As discussed earlier, in the section "Data Ingestion," we are dependent on a reliable data flow and ingestion process. To control that, we must ensure that we control the data latency. This is specified as the time difference between the time that the actual measurement was taken and the time at which it has been loaded.

Effectively leveraging data exploration can enhance the overall understanding of the characteristics of the data domain, thereby allowing for the creation of more accurate models.

Data exploration provides a set of simple tools to obtain some basic understanding of the data. The results of data exploration can be extremely powerful in grasping the structure of the data, the distribution of the values, and the presence of extreme values and interrelationships within the data set: this information provides guidance on applying the right kind of further data pre-processing and feature engineering (Lazzeri 2019b).

Once the raw data has been ingested and has been securely stored and explored, it is ready to be processed. The data preparation phase is basically taking the raw data and converting (transforming, reshaping) it into a form for the modeling phase. That may include simple operations such as using the raw data column as is with its actual measured value, standardized values, and more complex operations such as time lagging. The newly created data columns are referred to as data features, and the process of generating these is referred to as feature engineering. By the end of this process, we would have a new data set that has been derived from the raw data and can be used for modeling.

Data Pre-processing and Feature Engineering

Data pre-processing and feature engineering is the step where data scientists clean data sets from outliers and missing data and create additional features with the raw data to feed their machine learning models. Specifically, feature engineering is the process of transforming data into features to act as inputs for machine learning models such that good-quality features help in improving the overall model performance. Features are also very much dependent on the underlying problem we are trying to solve with a machine learning solution.

In machine learning, a feature is a quantifiable variable of the phenomenon you are trying to analyze, typically depicted by a column in a data set. Considering a generic two-dimensional data set, each observation is depicted by a row and each feature by a column, which will have a specific value for an observation, as illustrated in Figure 2.4.

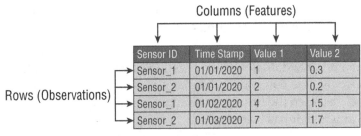

Figure 2.4: Understanding time series features

Thus, like the example in Figure 2.4, each row typically indicates a feature vector and the entire set of features across all the observations forms a two-dimensional feature matrix also known as a feature set. This is similar to data frames or spreadsheets representing two-dimensional data. Typically, machine learning algorithms work with these numeric matrices or tensors and hence most feature engineering techniques deal with converting raw data into some numeric representations that can be easily understood by these algorithms.

Features can be of two major types based on the data set:

- *Inherent raw features* are usually already your data set, with no need of extra data manipulation or feature engineering, as they are usually originated and collected directly from the data source.

- *Derived features* are usually created through data manipulation or feature engineering; in this stage, data scientists need to extract features from existing data attributes.

The following list includes some of the common data derived features that are included in the forecasting models:

- *Time driven features*: These features are derived from the date/time stamp data. They are extracted and converted into categorical features like these:

  - Time of day – This is the hour of the day, which takes values from 0 to 23.

  - Day of week – This represents the day of the week and takes values from 1 (Sunday) to 7 (Saturday).

  - Day of month – This represents the actual date and can take values from 1 to 31.

  - Month of year – This represents the month and takes values from 1 (January) to 12 (December).

  - Weekend – This is a binary value feature that takes the values of 0 for weekdays or 1 for weekend.

  - Holiday - This is a binary value feature that takes the values of 0 for a regular day or 1 for a holiday.

  - Fourier terms – The Fourier terms are weights that are derived from the time stamp and are used to capture the seasonality (cycles) in the data. Since we may have multiple seasons in our data, we may need multiple Fourier terms. For example, demand values may have yearly, weekly, and daily seasons/cycles, which will result in three Fourier terms.

- *Independent measurement features*: The independent features include all the data elements that we would like to use as predictors in our model. Here we exclude the dependent feature that we would need to predict.

- Lag feature – These are time-shifted values of the actual demand. For example, lag 1 features will hold the demand value in the previous hour (assuming hourly data) relative to the current time stamp. Similarly, we can add lag 2, lag 3, and so on. The actual combination of lag features that are used are determined during the modeling phase by evaluation of the model results.

- Long-term trending – This feature represents the linear growth in demand between years.

- *Dependent feature*: The dependent feature is the data column that we would like our model to predict. With supervised machine learning, we need to first train the model using the dependent features (which is also referred to as labels). This allows the model to learn the patterns in the data associated with the dependent feature. In energy demand forecast, we typically want to predict the actual demand and therefore we would use it as the dependent feature.

As we discussed in this section, data pre-processing and feature engineering are important steps to prepare, clean your data, and create features that will make your machine learning algorithms work better and produce more accurate results.

In the next section we will discuss how you can build forecasting models with your pre-processed data and features. We will take a closer look at different concepts and techniques that will help you generate multiple machine learning models for your time series data.

Modeling Building and Selection

The modeling phase is where the conversion of the data into a model takes place. In the core of this process there are advanced algorithms that scan the historical data (training data), extract patterns, and build a model. That model can be later used to predict on new data that has not been used to build the model.

Once we have a working reliable model, we can then use it to score new data that is structured to include the required features. The scoring process will make use of the persisted model (object from the training phase) and predict the target variable.

At this point, it is also important to distinguish between train, validation, and test sets in machine learning, as illustrated in Figure 2.5.

Figure 2.5: A representation of data set splits

▪ *Train data set* – The train data set represents the amount of data that data scientists decide to use to fit machine learning models. Through the train data set, a machine learning algorithm can be trained to learn from historical data in order to predict future data points.

▪ *Validation data set* – The validation data set is the amount of data used to offer an unbiased evaluation of a model fit on the train data set, while tuning model hyperparameters. Data scientists usually leverage validation data sets to fine-tune a machine learning model hyperparameters. Hyperparameters are additional factors in the model, whose values are used to control, and eventually improve, the learning process of the model itself. Data scientists observe the validation set results to update the levels of hyperparameters and, consequently, improve their models.

▪ *Test data set* – The test data set is the amount of data used to determine whether a model is underfitting (the model is performing poorly on the train data set) or overfitting (the model is performing well on the train data set but not on the test data set) the train data by looking at the prediction error on the train and test data sets. The test data set is only used once a model is completely trained and validated through the train and validation data sets. Data scientists usually leverage the test data set to evaluate different machine learning models.

Now that you understand better the difference between these data sets, it is also important to look for recommendations on how to split our data set into train, validation, and test sets. This decision mainly depends on two factors:

▪ The total number of samples in our data

▪ The actual model you are training.

Some models need substantial data to train upon, so in this case we would need to optimize for the larger training sets. Models with very few hyperparameters will be easy to validate and tune, so we can probably reduce the size of our validation set, but if our model has many hyperparameters, we would want to have a large validation set as well. Also, if we have to deal with a model with no hyperparameters or ones that cannot be easily tuned, we probably won't need a validation data set. All in all, like many other things in machine learning, the train-test-validation split ratio is also quite specific to your use case (Lazzeri 2019b).

In the following sections, we will discuss some of the most popular classical time series techniques and deep learning methods and how they can be applied to your data to build, train, test, and validate your forecasting models. Some of these techniques will be discussed in more detail in Chapter 4 and Chapter 5.

An Overview of Demand Forecasting Modeling Techniques

The ability to accurately forecast a sequence into the future is critical in many industries: finance, supply chain, and manufacturing are just a few examples. Classical time series techniques have served this task for decades, but now deep learning methods—similar to those used in computer vision and automatic translation—have the potential to revolutionize time series forecasting as well (Lazzeri 2019b).

In the case of demand forecasting, we make use of historical data, which is ordered by time. We generally refer to data that includes the time dimension as time series. The goal in time series modeling is to find time-related trends, seasonality, and autocorrelation (correlation over time) and formulate those into a model.

In recent years advanced algorithms have been developed to accommodate time series forecasting and to improve forecasting accuracy. I briefly discuss a few of them here. This information is not intended to be used as a machine learning and forecasting overview but rather as a short survey of modeling techniques that are commonly used for demand forecasting.

- *Moving average (MA)* – This is one of the first analytical techniques that has been used for time series forecasting, and it is still one of the most commonly used techniques as of today. It is also the foundation for more advanced forecasting techniques. With moving average, we are forecasting the next data point by averaging over the K most recent points, where K denotes the order of the moving average. The moving average technique has the effect of smoothing the forecast and therefore may not handle well large volatility in the data.

- *Exponential smoothing* – This is a family of various methods that use weighted average of recent data points to predict the next data point. The idea is to assign higher weights to more recent values and gradually decrease this weight for older measured values. There are a number of different methods within this family, including handling of seasonality in the data such as the Holt-Winters Seasonal Method. Some of these methods also factor in the seasonality of the data.

- *Autoregressive Integrated Moving Average (ARIMA)* – This is another family of methods that is commonly used for time series forecasting. It practically combines autoregression methods with moving average. Autoregression methods use regression models by taking previous time series values in order to compute the next date point. ARIMA methods also apply differencing methods that include calculating the difference between data points and using those instead of the original measured value. Finally,

ARIMA also makes use of the moving average techniques that are discussed above. The combination of all of these methods in various ways is what constructs the family of ARIMA methods. ETS and ARIMA are widely used today for demand forecasting and many other forecasting problems. In many cases, these are combined together to deliver very accurate results.

■ *General multiple regression* – This could be the most important modeling approach within the domain of machine learning and statistics. In the context of time series, we use regression to predict the future values (for example, of demand). In regression, we take a linear combination of the predictors and learn the weights (also referred to as coefficients) of those predictors during the training process. The goal is to produce a regression line that will forecast our predicted value. Regression methods are suitable when the target variable is numeric and therefore also fits time series forecasting. There is a wide range of regression methods, including very simple regression models such as linear regression and more advanced ones such as decision trees, random forests, neural networks, and boosted decision trees.

Due to their applicability to many real-life problems, such as fraud detection, spam email filtering, finance, and medical diagnosis, and their ability to produce actionable results, deep learning neural networks have gained a lot of attention in recent years. Generally, deep learning methods have been developed and applied to univariate time series forecasting scenarios, where the time series consists of single observations recorded sequentially over equal time increments (Lazzeri 2019a).

For this reason, they have often performed worse than naïve and classical forecasting methods, such as exponential smoothing and ARIMA. This has led to a general misconception that deep learning models are inefficient in time series forecasting scenarios, and many data scientists wonder whether it's really necessary to add another class of methods, like convolutional neural networks or recurrent neural networks, to their time series toolkit. Deep learning is a subset of machine learning algorithms that learn to extract these features by representing input data as vectors and transforming them with a series of clever linear algebra operations into a given output (Lazzeri 2019a).

Data scientists then evaluate whether the output is what they expected using an equation called loss function. The goal of the process is to use the result of the loss function from each training input to guide the model to extract features that will result in a lower loss value on the next pass. This process has been used to cluster and classify large volumes of information, like millions of satellite images; thousands of video and audio recordings from YouTube, and historical, textual, and sentiment data from Twitter.

Deep learning neural networks have three main intrinsic capabilities:

- They can learn from arbitrary mappings from inputs to outputs.
- They support multiple inputs and outputs.
- They can automatically extract patterns in input data that spans over long sequences.

Thanks to these three characteristics, they can offer a lot of help when data scientists deal with more complex but still very common problems, such as time series forecasting. After having experimented with different time series forecasting models, you need to select the best model based on your data and specific time series scenario. Model selection is an integral part of the model development process, in the next section we will discuss how model evaluation can help you find the best model that represents your data and understand how well the chosen model will work in the future (Lazzeri 2019b).

Model Evaluation

Model evaluation has a critical role within the modeling step. At this step we look into validating the model and its performance with real-life data. During the modeling step, we use a part of the available data for training the model. During the evaluation phase, we take the remainder of the data to test the model. Practically it means that we are feeding the model new data that has been restructured and contains the same features as the training data set.

However, during the validation process, we use the model to predict the target variable rather than provide the available target variable. We often refer to this process as model scoring. We would then use the true target values and compare them with the predicted ones. The goal here is to measure and minimize the prediction error, meaning the difference between the predictions and the true value.

Quantifying the error measurement is key since we would like to fine-tune the model and validate whether the error is actually decreasing. Fine-tuning the model can be done by modifying model parameters that control the learning process or by adding or removing data features (referred to as parameters sweep). Practically that means that we may need to iterate between the feature engineering, modeling, and model evaluation phases multiple times until we are able to reduce the error to the required level.

It is important to emphasize that the prediction error will never be zero as there is never a model that can perfectly predict every outcome. However, there is a certain magnitude of error that is acceptable by the business. During the validation process, we would like to ensure that our model prediction error is at or better than the business tolerance level. It is therefore important to set the level of the tolerable error at the beginning of the cycle during the problem formulation phase.

There are various ways in which prediction error can be measured and quantified. Specifically, there is an evaluation technique relevant to time series and in specific for demand forecast: the MAPE. MAPE stands for mean absolute percentage error. With MAPE we are computing the difference between each forecasted point and the actual value of that point. We then quantify the error per point by calculating the proportion of the difference relative to the actual value. At the last step we average these values. The mathematical formula used for MAPE is as follows:

$$MAPE = \left(\frac{1}{n} \Sigma \frac{\left| Actual\ Values - Forecasted\ Values \right|}{\left| Actual\ Values \right|} \right) \times 100$$

The MAPE is scale sensitive and should not be used when working with low-volume data. Notice that because *Actual Values* also represents the denominator of the equation, the MAPE is undefined when actual demand is zero. Furthermore, when the actual value is not zero, but quite small, the MAPE will often take on extreme values. This scale sensitivity renders the MAPE close to worthless as an error measure for low-volume data.

There are a few other evaluation techniques that are important to mention here:

- *The mean absolute deviation (MAD)* – This formula measures the size of the error in units. The MAD is a good statistic to use when analyzing the error for a single item. However, if you aggregate MADs over multiple items, you need to be careful about high-volume products dominating the results—more on this later. The MAPE and the MAD are by far the most commonly used error measurement statistics. There are a slew of alternative statistics in the forecasting literature, many of which are variations on the MAPE and the MAD (Stellwagen 2011).

- *The mean absolute deviation (MAD)/mean ratio* – This is an alternative to the MAPE that is better suited to intermittent and low-volume data. As stated previously, percentage errors cannot be calculated when the actual equals zero, and they can take on extreme values when dealing with low-volume data. These issues become magnified when you start to average MAPEs over multiple time series. The MAD/mean ratio tries to overcome this problem by dividing the MAD by the mean—essentially rescaling the error to make it comparable across time series of varying scales. The statistic is calculated exactly as the name suggests—it is simply the MAD divided by the mean (Stellwagen 2011).

- *The Geometric Mean Relative Absolute Error (GMRAE)* – This metric is used to measure out-of-sample forecast performance. It is calculated using the relative error between the naïve model (i.e., next period's forecast is this period's actual) and the currently selected model. A GMRAE of 0.54 indicates that the size of the current model's error is only 54 percent of the size of the error generated using the naïve model for the same data set. Because the GMRAE is based on a relative error, it is less scale sensitive than the MAPE and the MAD (Stellwagen 2011).

- *The symmetric mean absolute percentage error (SMAPE)* – This is a variation on the MAPE that is calculated using the average of the absolute value of the actual and the absolute value of the forecast in the denominator. This statistic is preferred to the MAPE by some and was used as an accuracy measure in several forecasting competitions (Stellwagen 2011).

After selecting the best model for their forecasting solution, data scientists usually need to deploy it. In the next section we will take a closer look at the deployment process, that is, the method by which you integrate a machine learning model into an existing production environment in order to start using it to make practical business decisions based on data. It is one of the last stages in the machine learning life cycle.

Model Deployment

Model deployment is the method by which you integrate a machine learning model into an existing production environment in order to start using it to make practical business decisions based on data. It is one of the last stages in the machine learning life cycle and can be one of the most cumbersome. Often, an organization's IT systems are incompatible with traditional model-building languages, forcing data scientists and programmers to spend valuable time and brainpower rewriting them (Lazzeri 2019c).

Once we have nailed down the modeling phase and validated the model performance, we are ready to go into the deployment phase. In this context, deployment means enabling the customer to consume the model by running actual predictions on it at large scale.

Machine learning model deployment is the process by which a machine learning algorithm is converted into a web service. We refer to this conversion process as operationalization: to operationalize a machine learning model means to transform it into a consumable service and embed it into an existing production environment (Lazzeri 2019c).

Model deployment is a fundamental step of the machine learning model workflow (Figure 2.6) because, through machine learning model deployment, companies can begin to take full advantage of the predictive and intelligent models they build, develop business practices based on their model results, and therefore, transform themselves into actual data-driven businesses.

Figure 2.6: Machine learning model workflow

When we think about machine learning, we focus our attention on key components of the machine learning workflow, such as data sources and ingestion, data pipelines, machine learning model training and testing, how to engineer new features, and which variables to use to make the models more accurate. All these steps are important; however, thinking about how we are going to consume those models and data over time is also a critical step in every machine learning pipeline. We can only begin extracting real value and business benefits from a model's predictions when it has been deployed and operationalized (Lazzeri 2019c).

Successful model deployment is fundamental for data-driven enterprises for the following key reasons:

- Deployment of machine learning models means making models available to external customers and/or other teams and stakeholders in your company.

- By deploying models, other teams in your company can use them, send data to them, and get their predictions, which are in turn populated back into the company systems to increase training data quality and quantity.

- Once this process is initiated, companies will start building and deploying higher numbers of machine learning models in production and master robust and repeatable ways to move models from development environments into business operations systems.

Many companies see machine-enablement effort as a technical practice. However, it is more of a business-driven initiative that starts within the company; in order to become a data-driven company, it is important that the people, who today successfully operate and understand the business, collaborate closely with those teams that are responsible for the machine learning deployment workflow (Lazzeri 2019c).

Right from the first day of a machine learning solution creation, data science teams should interact with business counterparts. It is very essential to maintain constant interaction to understand the model *experimentation* process parallel to the model *deployment* and *consumption* steps. Most organizations struggle to unlock machine learning's potential to optimize their operational processes and get data scientists, analysts, and business teams speaking the same language.

Moreover, machine learning models must be trained on historical data, which demands the creation of a prediction data pipeline, an activity requiring multiple tasks including data processing, feature engineering, and tuning. Each task, down to versions of libraries and handling of missing values, must be exactly duplicated from the development to the production environment. Sometimes, differences in technology used in development and in production contribute to difficulties in deploying machine learning models.

Companies can use machine learning pipelines to create and manage workflows that stitch together machine learning phases. For example, a pipeline might include data preparation, model training, model deployment, and inference/scoring phases. Each phase can encompass multiple steps, each of which can run unattended in various compute targets. Pipeline steps are reusable and can be run without rerunning subsequent steps if the output of that step has not changed. Pipelines also allow data scientists to collaborate while working on separate areas of a machine learning workflow (Lazzeri 2019c).

Building, training, testing, and finally, deploying machine learning models is often a tedious and slow process for companies that are looking at transforming their operations with machine learning. Moreover, even after months of development, which delivers a machine learning model based on a single algorithm, the management team has little means of knowing whether their data scientists have created a great model and how to scale and operationalize it.

Next, I share a few guidelines on how a company can select the right tools to succeed with model deployment. I will illustrate this workflow using the Azure Machine Learning service, but it can be also used with any machine learning product of your choice.

The model deployment workflow should be based on the following three simple steps (Lazzeri 2019c):

- Register the model – A registered model is a logical container for one or more files that make up your model. For example, if you have a model that is stored in multiple files, you can register them as a single model in the workspace. After registration, you can then download or deploy the registered model and receive all the files that were registered.
  Machine learning models are registered when you create an Azure Machine Learning workspace. The model can come from Azure Machine Learning or from somewhere else.

- Prepare to deploy (specify assets, usage, compute target) – To deploy a model as a web service, you must create an inference configuration (`inference config`) and a deployment configuration. Inference, or model scoring, is the phase where the deployed model is used for prediction, most commonly on production data. In the inference config, you specify the scripts and dependencies needed to serve your model. In the deployment config, you specify details of how to serve the model on the compute target.

The entry script receives data submitted to a deployed web service and passes it to the model. It then takes the response returned by the model and returns that to the client. The script is specific to your model; it must understand the data that the model expects and returns (Lazzeri 2019c). The script contains two functions that load and run the model:

- `init()` – Typically, this function loads the model into a global object. This function is run only once when the Docker container for your web service is started.

- `run(input _ data)` – This function uses the model to predict a value based on the input data. Inputs and outputs to the run typically use JSON for serialization and de-serialization. You can also work with raw binary data. You can transform the data before sending it to the model or before returning it to the client.

When you register a model, you provide a model name used for managing the model in the Azure Registry. You use this name with the `Model.get _ model _ path()` to retrieve the path of the model file(s) on the local file system. If you register a folder or a collection of files, this API returns the path to the directory that contains those files.

- Deploy the model to the compute target – Finally, before deploying, you must define the deployment configuration. The deployment configuration is specific to the compute target that will host the web service. For example, when deploying locally, you must specify the port where the service accepts requests. Table 2.1 lists the compute targets, or compute resources, that can be used to host your web service deployment.

Table 2.1: Examples of compute targets that can be used to host your web service deployment

COMPUTE TARGET	USAGE	DESCRIPTION
Local web service	Testing/ debug	Good for limited testing and troubleshooting. Hardware acceleration depends on using libraries in the local system.
Notebook VM web service	Testing/ debug	Good for limited testing and troubleshooting.
Azure Kubernetes Service (AKS)	Real-time inference	Good for high-scale production deployments. Provides fast response time and autoscaling of the deployed service. Cluster autoscaling is not supported through the Azure Machine Learning SDK. To change the nodes in the AKS cluster, use the UI for your AKS cluster in the Azure portal.
Azure Container Instances (ACI)	Testing or dev	Good for low-scale, CPU-based workloads requiring <48 GB RAM.

Continues

Table 2.1 (*continued*)

COMPUTE TARGET	USAGE	DESCRIPTION
Azure Machine Learning Compute	Batch inference	Run batch scoring on serverless compute. Supports normal- and low-priority VMs.
Azure IoT Edge	IoT module	Deploy and serve ML models on IoT devices.
Azure Data Box Edge	via IoT Edge	Deploy and serve ML models on IoT devices.

When deploying a demand forecasting solution, we are interested in deploying an end-to-end solution that goes beyond the prediction web service and facilitates the entire data flow. At the time we invoke a new forecast, we would need to make sure that the model is fed with the up-to-date data features. That implies that the newly collected raw data is constantly ingested, processed, and transformed into the required feature set on which the model was built.

At the same time, we would like to make the forecasted data available for the end consuming clients. These are the steps that take place as part of the energy demand forecast cycle:

- Millions of deployed data meters are constantly generating power consumption data in real time.

- This data is being collected and uploaded into a cloud repository.

- Before being processed, the raw data is aggregated to a substation or regional level as defined by the business.

- The feature processing then takes place and produces the data that is required for model training or scoring—the feature set data is stored in a database.

- The retraining service is invoked to retrain the forecasting model–that updated version of the model is persisted so that it can be used by the scoring web service.

- The scoring web service is invoked on a schedule that fits the required forecast frequency.

- The forecasted data is stored in a database that can be accessed by the end consumption client.

- The consumption client retrieves the forecasts, applies it back into the grid, and consumes it in accordance with the required use case.

As illustrated in Figure 2.7, after we have built a set of models that perform well, we can operationalize them for other applications to consume.

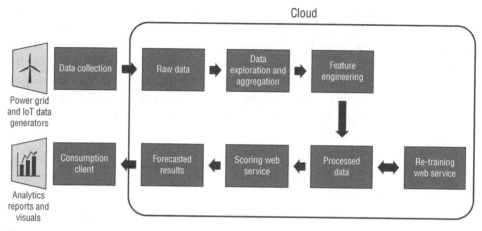

Figure 2.7: Energy demand forecast end-to-end solution

Depending on the business requirements, predictions are made either in real time or on a batch basis. To deploy models, you expose them with an open API. The interface enables the model to be easily consumed from various applications, such as these:

- Online websites
- Spreadsheets
- Dashboards
- Line-of-business applications
- Backend applications

Once the machine learning solution is deployed, it is crucial to finalize the project deliverables. In the next section, we will discuss how data scientists can confirm that the pipeline, the model, and their deployment in a production environment satisfy the customers' and the final users' objectives.

Forecasting Solution Acceptance

In the last stage of the time series forecasting solution development, data scientists need to confirm that the pipeline, the model, and their deployment in a production environment satisfy the customers' and the final users' objectives.

Assuming that an organization has all the right ingredients, including the right employee culture, they still need to have the right technology platform in place to support data scientists' productivity and helps them innovate and iterate rapidly. A modern cloud analytics environment will make it super easy to collect data, analyze, experiment, and quickly put things into production with a targeted set of customers. This sort of capability is becoming a must-have for data-driven organizations, large and small (Lazzeri 2019c).

Without such a platform, data scientists will find it hard to rapidly iterate over many experiments, learn quickly from their failures and successes, and discover interesting actionable insights from data. Without the right culture, infrastructure, and tools, an organization will eventually find itself lagging behind nimbler competitors.

There are two main tasks addressed in the customer acceptance stage:

- *System validation* – Confirm that the deployed model and pipeline meet the customer's needs.

- *Project handoff* – Hand the project off to the entity that is going to run the system in production (Lazzeri 2019b).

The customer should validate that the system meets their business needs and that it answers the questions with acceptable accuracy to deploy the system to production for use by their client's application. All the documentation is finalized and reviewed. The project is handed off to the entity responsible for operations. This entity might be, for example, an IT or customer data-science team or an agent of the customer that's responsible for running the system in production.

The ability to successfully navigate through this documentation could very well determine the viability and longevity of a forecasting solution based on machine learning.

To make these data science life cycles more concrete, I will now introduce a demand forecasting use case that will be used throughout this book for explanatory reasons and to showcase a few forecasting scenarios and sample codes.

Use Case: Demand Forecasting

In this final section, I introduce a real-world data science scenario that I will use throughout this book to showcase some of the concepts, steps, and techniques discussed so far: demand forecasting use case. I believe everyone must learn to smartly work with huge amounts of data, hence large data sets are included, open and free to access.

In the past few years, Internet of Things (IoT), alternative energy sources, and big data have merged to create vast opportunities in the utility and energy domain. At the same time, the utility and the entire energy sector have seen consumption flattening out with consumers demanding better ways to control their use of energy. Hence, the utility and smart grid companies are in great need to innovate and renew themselves. Furthermore, many power and utility grids are becoming outdated and very costly to maintain and manage.

Within the energy sector, there could be many ways in which demand forecasting can help solve critical business problems. In fact, demand forecasting can be considered the foundation for many core use cases in the industry. In general, we consider two types of energy demand forecasts: short term and long term. Each one may serve a different purpose and utilize a different approach. The main difference between the two is the forecasting horizon, meaning the range of time into the future for which we would forecast.

Within the context of energy demand, short-term load forecasting (STLF) is defined as the aggregated load that is forecasted in the near future on various parts of the grid (or the grid as a whole). In this context, *short term* is defined to be a time horizon within the range of 1 hour to 24 hours. In some cases, a horizon of 48 hours is also possible. Therefore, STLF is very common in an operational use case of the grid.

STLF models are mostly based on the near past (last day or week) consumption data and use forecasted temperature as an important predictor. Obtaining accurate temperature forecasts for the next hour and up to 24 hours is becoming less of a challenge nowadays. These models are less sensitive to seasonal patterns or long-term consumption trends.

SLTF solutions are also likely to generate a high volume of prediction calls (service requests) since they are being invoked on an hourly basis and in some cases even with higher frequency. It is also very common to see implantation, where each individual substation or transformer is represented as a stand-alone model and therefore the volume of prediction requests is even greater.

The goal of long-term load forecasting (LTLF) is to forecast power demand with a time horizon ranging from one week to multiple months (and in some cases for a number of years). This range of horizon is mostly applicable for planning and investment use cases.

For long-term scenarios, it is important to have high-quality data that covers a span of multiple years (minimum three years). These models will typically extract seasonality patterns from the historical data and make use of external predicators such as weather and climate patterns.

It is important to clarify that the longer the forecasting horizon is, the less accurate the forecast may be. It is therefore important to produce some confidence intervals along with the actual forecast that would allow humans to factor the possible variation into their planning process.

Since the consumption scenario for LTLF is mostly planning, we can expect much lower prediction volumes (as compared to STLF). We would typically see these predictions embedded into visualization tools such as Excel or PowerBI and be invoked manually by the user.

Table 2.2 compares STLF and LTLF in respect to the most important attributes.

Table 2.2: Short-term versus long-term predictions

ATTRIBUTE	SHORT-TERM LOAD FORECAST	LONG-TERM LOAD FORECAST
Forecast horizon	From 1 hour to 48 hours	From 1 to 6 months or more
Data granularity	Hourly	Hourly or daily
Typical use cases	Demand/supply balancing	Long-term planning
	Pick hour forecasting	Grid assets planning
	Demand response	Resource planning
Typical predictors	Day or week	Month of year
	Hour of day	Day of month
	Hourly temperature	Long-term temperature and climate
Historical data range	Two to three years' worth of data	Five to 10 years' worth of data
Typical accuracy	MAPE (mean absolute percentage error) of 5% or lower	MAPE (mean absolute percentage error) of 25% or lower
Forecast frequency	Produced every hour or every 24 hours	Produced once monthly, quarterly, or yearly

As can be seen from Table 2.2, it is quite important to distinguish between the short- and the long-term forecasting scenarios as these represent different business needs and may have different deployment and consumption patterns.

Any advanced analytics-based solution relies on data. Specifically, when it comes to predictive analytics and forecasting, we rely on ongoing, dynamic flow of data. In the case of energy demand forecasting, this data can be sourced directly from smart meters or be already aggregated on an on-premises database. We also rely on other external sources of data, such as weather and temperature. This ongoing flow of data must be orchestrated, scheduled, and stored.

For this specific use case, we will be using the public data set "Load Forecasting Data" from the GEFCom2014 competition. The load forecasting track of GEFCom2014 was about probabilistic load forecasting. The complete data was published as the appendix of our GEFCom2014 paper: (Hong et al. 2016).

For the rest of this book, you can download the data set from `aka.ms/EnergyDataSet`. This data set consists of three years of hourly electricity load and temperature values between 2012 and 2014. An accurate and fast-performing prediction requires implementation of three predictive models:

- Long-term model that enables forecasting of power consumption during the next few weeks or months

- Short-term model that enables prediction of overload situation during the next hour

- Temperature model that provides forecasting of future temperature over multiple scenarios

As temperature is an important predictor for the long-term model, there is a need to constantly produce multi-scenario temperature forecasts and feed them as input into to the long-term model. Moreover, to gain higher prediction accuracy in the short term, a more granular model is dedicated for each hour of the day.

After identifying the required data sources, we would like to ensure that raw data that has been collected includes the correct data features. To build a reliable demand forecast model, we would need to ensure that the data collected includes data elements that can help predict the future demand. Here are some basic requirements concerning the data structure (schema) of the raw data.

The raw data consists of rows and columns. Each measurement is represented as a single row of data. Each row of data includes multiple columns (also referred to as features or fields):

- *Time stamp* – The time stamp field represents the actual time the measurement was recorded. It should comply with one of the common date/time formats. Both date and time parts should be included. In most cases, there is no need for the time to be recorded till the second level of granularity. It is important to specify the time zone in which the data is recorded.

- *Load value* – This is the actual consumption at a given date/time. The consumption can be measured in kWh (kilowatt-hour) or any other preferred units. It is important to note that the measurement unit must stay consistent across all measurements in the data. In some cases, consumption can be supplied over three power phases. In that case we would need to collect all the independent consumption phases.

- *Temperature* – The temperature is typically collected from an independent source. However, it should be compatible with the consumption data. It should include a time stamp as described above that will allow it to be synchronized with the actual consumption data. The temperature value can be specified in degrees Celsius or Fahrenheit but should stay consistent across all measurements.

We will be using this demand forecasting use case to discuss many classical, machine learning, and deep learning approaches in the next few chapters of this book.

Conclusion

The purpose of this second chapter was to provide an end-to-end systematic template for time series forecasting from a practical perspective: I have introduced some of the most important concepts to build your end-to-end time series forecasting solutions.

In particular, we discussed the following concepts in detail:

- ▪ *Time series forecasting template* – This is a set of tasks that leads from defining a business problem through to the outcome of having a time series forecast model deployed and ready to be consumed externally or across your company.

Our template is based on the following steps:

- ▪ *Business understanding and performance metrics definition* – The business understanding performance metrics definition step outlines the business aspect we need to understand and consider prior to making an investment decision.

- ▪ *Data ingestion* – Data ingestion is the process of collecting and importing data to be cleaned and analyzed or stored in a database.

- ▪ *Data exploration and understanding* – Once the raw data has been ingested and securely stored, it is ready to be explored and understood.

- ▪ *Data pre-processing and feature engineering* – Data pre-processing and feature engineering is the step where data scientists clean data sets from outliers and missing data and create additional features with the raw data to feed their machine learning models.

- ▪ *Modeling building and selection* – The modeling phase is where the conversion of the data into a model takes place. In the core of this process there are advanced algorithms that scan the historical data (training data), extract patterns, and build a model. That model can be later used to predict on new data that has not been used to build the model.

- ▪ *Model deployment* – Deployment is the method by which you integrate a machine learning model into an existing production environment in order to start using it to make practical business decisions based on data. It is one of the last stages in the machine learning life cycle.

■ *Forecasting solution acceptance* – In the last stage of the time series fore-casting solution deployment, data scientists need to confirm and validate that the pipeline, the model, and their deployment in a production environment satisfy their success criteria.

■ *Use case: demand forecasting* – At the end of this chapter, I introduced a real-world data science scenario that we will use throughout this book to showcase some of the time series concepts, steps, and techniques discussed.

In the next chapter, "Time Series Data Preparation," we will discuss some of the most popular concepts and techniques for time series data preparation. In particular, we will take a closer look at important steps when handing time series data:

■ Python for time series data

■ Time series data exploration and understanding

■ Time series feature engineering

Time Series Data Preparation

In this chapter, I will walk you through the most important steps to prepare your time series data for forecasting models. *Data preparation* is the practice of transforming your raw data so that data scientists can run it through machine learning algorithms to discover insights and, eventually, make predictions.

Each machine learning algorithm expects data as input that needs to be formatted in a very specific way, so time series data sets generally require some cleaning and feature engineering processes before they can generate useful insights. Time series data sets may have values that are missing or may contains outliers, hence the need for the data preparation and cleaning phase is essential. Since the time series data has temporal property, only some of the statistical methodologies are appropriate for time series data. Good time series data preparation produces clean and well-curated data, which leads to more practical, accurate predictions.

Specifically, in this chapter we will discuss the following:

■ *Python for Time Series Data* – Python is a very powerful programming language to handle data, offering an assorted suite of libraries for time series data and excellent support for time series analysis. In this section of Chapter 3, you will see how libraries such as SciPy, NumPy, Matplotlib, pandas, statsmodels, and scikit-learn can help you prepare, explore, and analyze your time series data.

- *Time Series Exploration and Understanding* – In this section of Chapter 3, you will learn the first steps to take to explore, analyze, and understand time series data. Specifically, this section will focus on how to get started with time series data analysis and how to calculate and review summary statistics for time series data; then you will learn how to perform data cleaning of missing periods in a time series; finally, you will see how to perform time series data normalization and standardization.

- *Time Series Feature Engineering* – Feature engineering is the process of using raw data to create additional features or variables to augment your data set. In this section of Chapter 3, you will learn how to perform feature engineering on time series data with two goals in mind: preparing the proper input time series data set, compatible with the machine learning algorithm requirements; and improving the performance of machine learning models. If feature engineering is done correctly, it creates a stronger relationships between new input features and the output feature for the supervised learning algorithm to model.

Let's get started and learn how you can easily use Python and leverage multiple libraries to prepare and analyze your time series data.

Python for Time Series Data

Python is currently one of the most dominant platforms for time series data because of the excellent library support. There are a few Python libraries for time series, as Figure 3.1 illustrates.

As shown in Figure 3.1, SciPy is a Python-based ecosystem of open-source software for mathematics, science, and engineering. It includes modules for statistics, optimization, integration, linear algebra, Fourier transforms, signal and image processing, ODE solvers, and more. You can find more information about SciPy at the following links:

- Website: `scipy.org/`
- Documentation: `docs.scipy.org/`
- Source code: `github.com/scipy/scipy`
- Bug reports: `github.com/scipy/scipy/issues`
- Code of Conduct: `scipy.github.io/devdocs/dev/conduct/code_of_conduct.html`

SciPy is built to work with NumPy arrays, and provides many user-friendly and efficient numerical routines, such as routines for numerical integration and optimization. They are easy to use, install, and run on all popular operating systems.

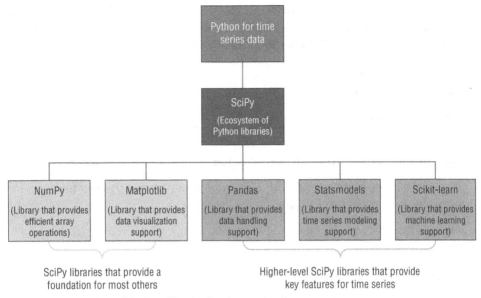

Figure 3.1: Overview of Python libraries for time series data

NumPy is the fundamental package for scientific computing with Python. It contains, among other things, the following components:

- A powerful N-dimensional array object
- Sophisticated (broadcasting) functions
- Tools for integrating C/C++ and Fortran code
- Useful linear algebra, Fourier transform, and random number capabilities

NumPy is a useful Python library that can also be used as multidimensional container of generic data. This additional capability allows NumPy to be efficient and fast with many different databases. You can find more information about SciPy at the following links:

- Website: `numpy.org`
- Documentation: `docs.scipy.org/`
- Source code: `github.com/numpy/numpy`
- Bug reports: `github.com/numpy/numpy/issues`
- Contributing: `numpy.org/devdocs/dev/index.html`

Matplotlib is a data visualization library built on NumPy arrays and designed to work with the broader SciPy stack. It is a complete library for plotting data, and it allows data scientists to build dynamic and interactive visualizations. Moreover, it allows data scientists to connect to and consume large amounts of data to build agile visualizations.

Matplotlib can be used in Python scripts, the Python and IPython shell, web application servers, and various graphical user interface toolkits. You can find more information about Matplotlib at the following links:

- Website: `matplotlib.org/`
- Documentation: `matplotlib.org/users/index.html`
- Source code: `github.com/matplotlib/matplotlib`
- Bug reports: `github.com/matplotlib/matplotlib/issues`
- Contributing: `matplotlib.org/devdocs/devel/contributing.html`

There are three higher-level SciPy libraries that provide the key features for time series forecasting in Python:

- pandas
- statsmodels
- scikit-learn

pandas is built on the NumPy package and it is a very handy and popular package among data scientists for importing and analyzing data. pandas's main data structure is the *DataFrame*, which is a table of data with rows and named columns. pandas offers time-series-specific functionalities, such as date range generation and frequency conversion, moving window statistics, date shifting, and lagging. You can find more information about pandas at the following links:

- Website: `pandas.pydata.org/`
- Documentation: `pandas.pydata.org/docs/user _ guide/index.html`
- Source code: `github.com/pandas-dev/pandas`
- Bug reports: `github.com/pandas-dev/pandas/issues`
- Contributing: `pandas.pydata.org/docs/development/index.html`

statsmodels is a Python module that supports a wide range of statistical functions and classes using pandas DataFrames. It is also used by data scientists for conducting statistical tests and statistical data exploration. You can find more information about statsmodels at the following links:

- Website: `statsmodels.org/`
- Documentation: `statsmodels.org/stable/user-guide.html`
- Source code: `github.com/statsmodels/statsmodels`
- Bug reports: `github.com/statsmodels/statsmodels/issues`
- Contributing: `github.com/statsmodels/statsmodels/blob/master/CONTRIBUTING.rst`

scikit-learn is library largely written in Python and built upon NumPy, SciPy, and Matplotlib. This library offers many useful capabilities for machine learning and statistical modeling, including vector machines, random forests, k-neighbors, classification, regression, and clustering. You can find more information about scikit-learn at the following links:

- Website: `scikit-learn.org/stable/`
- Documentation: `scikit-learn.org/stable/user _ guide.html`
- Source code: `github.com/scikit-learn/scikit-learn`
- Bug reports: `github.com/scikit-learn/scikit-learn/issues`
- Contributing: `scikit-learn.org/stable/developers/contributing.html`

In the rest of this chapter, we will take a closer look at how Numpy, Matplotlib, and pandas can be very helpful for dealing with time series data, while in the next chapters we will discuss how to use statsmodels and scikit-learn for time series modeling and machine learning respectively.

Common Data Preparation Operations for Time Series

pandas (`pandas.pydata.org`) offers a diverse portfolio of capabilities to support time series data. The following key features are important for time series forecasting in pandas:

- Parsing time series information from various sources and formats
- Generating sequences of fixed-frequency dates and time spans
- Manipulating and converting date times with time zone information
- Resampling or converting a time series to a particular frequency
- Performing date and time arithmetic with absolute or relative time increments

pandas captures four general time related concepts:

- *Date time*: A specific date and time with time zone support, such as month, day, year, hour, second, microsecond.
- *Time deltas*: An absolute time duration used for manipulating dates.
- *Time spans*: A duration of time defined by a point in time and its associated frequency.
- *Date offsets*: A relative time duration that respects calendar arithmetic.

As illustrated in Table 3.1, the four general time-related concepts have different characteristics and primary creation methods.

Table 3.1: Four general time-related concepts supported in pandas

CONCEPT	SCALAR CLASS	ARRAY CLASS	DATA TYPE	CREATION METHOD
Date times	Time stamp	DatetimeIndex	datetime64[ns] or datetime64[ns, tz]	to_datetime or date_range
Time deltas	Timedelta	TimedeltaIndex	timedelta64[ns]	to_timedelta or timedelta_range
Time spans	Period	PeriodIndex	period[freq]	Period or period_range
Date offsets	DateOffset	None	None	DateOffset

In Table 3.1, there are a few additional concepts that are important to explain for those new to Python:

- *Scalar class*: A scalar class is a numerical value alone and defines a vector space. A quantity described by multiple scalars, such as having both direction and magnitude, is called a vector.
- *Array class*: An array class is a data structure that holds a fixed number of values of the same data type.
- *Data type*: A data type is a labeling of different data objects. It represents the type of values in your data set used to understand how to store and manipulate data.
- *Primary creation method*: A group of functions and operations that are typically written and used to perform a single or a set of actions on your data. Functions, if properly written, provide better efficiency, reproducibility, and modularity to your Python code.

In the next few sections, we will discuss some of the most common operations that you can perform using pandas for time series data.

Time stamps vs. Periods

The pandas library in Python provides comprehensive and built-in support for time series data. The examples below and for the rest of the chapter focuse on energy demand forecasting. In general, demand forecasting is the practice of predicting future demand (or quantity of specific products and services) by using historical demand data to make informed business decisions about future plans, such as planning a product inventory optimization strategy, allocating marketing investments, and estimating the cost of new products and services to meet customer demand and needs.

Energy demand forecasting is a type of demand forecasting where the goal is to predict the future load (or energy demand) on an energy grid. It is a critical business operation for companies in the energy sector, as operators need to maintain the fine balance between the energy consumed on a grid and the energy supplied to it. Typically, grid operators can take short-term decisions to manage energy supply to the grid and keep the load in balance. An accurate short-term forecast of energy demand is therefore essential for the operator to make these decisions with confidence. This scenario details the construction of a machine learning energy demand forecasting solution.

Let's first look at the concept of *time stamped* data, which is the most essential type of time series data and helps you combine values with specific points in time. This means that each data point of your data set will have a temporal information that you can leverage, as shown in the sample code below:

```
# import necessary Python packages
import pandas as pd
import datetime as dt
import numpy as np

pd.Timestamp(dt.datetime(2014, 5, 1))

# or we can use
pd.Timestamp('2014-06-01')

# or we can use
pd.Timestamp(2014, 6, 1)
```

The result of these examples will be as follows:

```
Timestamp('2014-06-01 00:00:00')
```

In many time series scenarios, it can be also useful to link data points in your data set to a time interval instead. The time interval represented by `Period` can be inferred from datetime string format, as shown in the following example:

```
pd.Period('2014-06')
```

Output:

```
Period('2014-06', 'M')
```

Where 'M' stands for month. The time interval represented by `Period` can be also specified explicitly:

```
pd.Period('2014-06', freq='D')
```

Output:

```
Period('2014-06-01', 'D')
```

Where 'D' stands for day. Timestamp and Period can also be used as an index: in this case, lists of Timestamp and Period are automatically coerced to DatetimeIndex and PeriodIndex respectively. Let's start with the Timestamp, as the following example is showing:

```
dates = [pd.Timestamp('2014-06-01'),
         pd.Timestamp('2014-06-02'),
         pd.Timestamp('2014-06-03')]

ts_data = pd.Series(np.random.randn(3), dates)

type(ts_data.index)
```

Output:

```
pandas.core.indexes.datetimes.DatetimeIndex
```

If we want to look at the index format, we can run the following:

```
ts_data.index
```

Output:

```
DatetimeIndex(['2014-06-01', '2014-06-02', '2014-06-03'],
dtype='datetime64[ns]', freq=None)
```

Now let's look at the output of the time stamp:

```
ts_data
```

Output:

```
2014-06-01    -0.28
2014-06-02     0.41
2014-06-03    -0.53
dtype: float64
```

Let's now repeat the exercise, but with Period:

```
periods = [pd.Period('2014-01'), pd.Period('2014-02'), pd.Period
('2014-03')]
ts_data = pd.Series(np.random.randn(3), periods)
type(ts_data.index)
```

Output:

```
pandas.core.indexes.period.PeriodIndex
```

If we want to look at the index format, we can run the following:

```
ts_data.index
```

Output:

```
PeriodIndex(['2014-01', '2014-02', '2014-03'], dtype='period[M]',
freq='M')
```

Now let's look at the output of the time stamp:

```
ts_data
```

Output:

```
2014-01     0.43
2014-02    -0.10
2014-03    -0.04
Freq: M, dtype: float64
```

As you can see from the preceding examples, pandas allows you to capture both representations and convert between them. pandas represents time stamps using instances of `Timestamp` and sequences of time stamps using instances of `DatetimeIndex`.

Converting to Time stamps

Typically, data scientists represent the time component or the columns with dates in their data sets as the index of a Series or DataFrame, so that data preprocessing and cleaning manipulations can be performed with respect to the time element.

The example below shows that, when passed a Series, this returns a Series (with the same index):

```
pd.to_datetime(pd.Series(['Jul 31, 2012', '2012-01-10', None]))
```

The output will be

```
0    2012-07-31
1    2012-01-10
2           NaT
dtype: datetime64[ns]
```

On the other hand, the following example shows that, when passed a list-like, this is converted to a `DatetimeIndex`:

```
pd.to_datetime(['2012/11/23', '2012.12.31'])
```

The output will be

```
DatetimeIndex(['2012-11-23', '2012-12-31'], dtype='datetime64[ns]',
freq=None)
```

Finally, if you use dates that start with the day first (i.e. European style), you can pass the `dayfirst` flag:

```
pd.to_datetime(['04-01-2014 10:00'], dayfirst=True)
```

The output will be

```
DatetimeIndex(['2014-01-04 10:00:00'], dtype='datetime64[ns]',
freq=None)
```

The `dayfirst` flag works with multiple dates as well, as shown in the example below:

```
pd.to_datetime(['14-01-2014', '01-14-2012'], dayfirst=True)
```

The output will be

```
DatetimeIndex(['2014-01-14', '2012-01-14'], dtype='datetime64[ns]',
freq=None)
```

Providing a Format Argument

In addition to the required datetime string, a `format` argument can be passed to ensure specific parsing and define the structure and order of your time variable, as shown in the following example:

```
pd.to_datetime('2018/11/12', format='%Y/%m/%d')
```

The output will be

```
Timestamp('2018-11-12 00:00:00')
```

If you need to also define hours and minutes in your dates, you can format your time series data as follows:

```
pd.to_datetime('11-11-2018 00:00', format='%d-%m-%Y %H:%M')
```

The output will be

```
Timestamp('2018-11-11 00:00:00')
```

Moreover, date, datetime, and time objects all support a `strftime(format)` method to create a string representing the time under the control of an explicit format string. On the other hand, the `datetime.strptime()` class method creates a `datetime` object from a string representing a date and time and a corresponding format string.

In other words, `strptime()` and `strftime()` are two popular methods for converting objects from strings to datetime objects and the other way around. `strptime()` can read strings with date and time information and convert them to datetime objects, and `strftime()` converts datetime objects back into strings.

Table 3.2 provides a high-level comparison of `strftime()` versus `strptime()` methods.

Table 3.2: Comparison of `strftime()` and `strptime()` functionalities

	STRFTIME	STRPTIME
Functionality capability	Convert object to a string according to a given format	Parse a string into a datetime object given a corresponding format
Functionality method	Instance method	Class method
Python objects that support the functionality	`date`; `datetime`; `time`	`datetime`
Functionality signature	`strftime(format)`	`strptime(date_string, format)`

Indexing

pandas has a very useful method for data configuration and alignment called `reindex()`. It is a very handy technique, especially when data scientists have to deal with data sets relying on label-alignment functionality. In other words, *reindexing* your data sets means to adjust all data points to follow a given set of labels along a specific axis.

Moreover, `DatetimeIndex` can be used as an index for pandas objects containing time series structure. `DatetimeIndex` objects have the basic capabilities of regular `Index` objects and they also offer advanced time-series-specific methods for frequency processing, selection and slicing, as illustrated in the examples below, that use our ts_data set.

Let's first import the necessary Python packages to download the data set:

```
# import necessary Python packages to download the data set
import os
from common.utils import load_data
from common.extract_data import extract_data

# adjust the format of the data set
pd.options.display.float_format = '{:,.2f}'.format
np.set_printoptions(precision=2)
```

Then we can download our ts_data set:

```
# download ts_data set
# change the name of the directory with your folder name
data_dir = './energy'

if not os.path.exists(os.path.join(data_dir, 'energy.csv')):
    # download and move the zip file
    !wget https://mlftsfwp.blob.core.windows.net/mlftsfwp/GEFCom2014.zip
```

Continues

(continued)

```
!mv GEFCom2014.zip ./energy
# if not done already, extract zipped data and save as csv
extract_data(data_dir)
```

Finally, we need to load the data from CSV file into a pandas DataFrame. In the specific example below, we select and use only the load column of the ts_data set and we name it ts_data_load:

```
# load the data from csv into a pandas dataframe
ts_data_load = load_data(data_dir)[['load']]
ts_data_load.head()
```

DatetimeIndex can be used as an index for pandas objects containing time series structure, like the example below shows:

```
ts_data_load.index
```

The output will be

```
DatetimeIndex(['2012-01-01 00:00:00', '2012-01-01 01:00:00',
               '2012-01-01 02:00:00', '2012-01-01 03:00:00',
               '2012-01-01 04:00:00', '2012-01-01 05:00:00',
               '2012-01-01 06:00:00', '2012-01-01 07:00:00',
               '2012-01-01 08:00:00', '2012-01-01 09:00:00',
               ...
               '2014-12-31 14:00:00', '2014-12-31 15:00:00',
               '2014-12-31 16:00:00', '2014-12-31 17:00:00',
               '2014-12-31 18:00:00', '2014-12-31 19:00:00',
               '2014-12-31 20:00:00', '2014-12-31 21:00:00',
               '2014-12-31 22:00:00', '2014-12-31 23:00:00'],
              dtype='datetime64[ns]', length=26304, freq='H')
```

Now let's slice our ts_data set to access only specific parts of time sequences:

```
ts_data_load[:5].index
```

and

```
ts_data_load[::2].index
```

The output will respectively be

```
DatetimeIndex(['2012-01-01 00:00:00', '2012-01-01 01:00:00',
               '2012-01-01 02:00:00', '2012-01-01 03:00:00',
               '2012-01-01 04:00:00'],
              dtype='datetime64[ns]', freq='H')
```

and

```
DatetimeIndex(['2012-01-01 00:00:00', '2012-01-01 02:00:00',
               '2012-01-01 04:00:00', '2012-01-01 06:00:00',
               '2012-01-01 08:00:00', '2012-01-01 10:00:00',
               '2012-01-01 12:00:00', '2012-01-01 14:00:00',
               '2012-01-01 16:00:00', '2012-01-01 18:00:00',
               ...
               '2014-12-31 04:00:00', '2014-12-31 06:00:00',
               '2014-12-31 08:00:00', '2014-12-31 10:00:00',
               '2014-12-31 12:00:00', '2014-12-31 14:00:00',
               '2014-12-31 16:00:00', '2014-12-31 18:00:00',
               '2014-12-31 20:00:00', '2014-12-31 22:00:00'],
              dtype='datetime64[ns]', length=13152, freq='2H')
```

To provide convenience for accessing longer time series, you can also pass in the year or year and month as strings:

```
ts_data_load['2011-6-01']
```

The output will be

```
load
2012-06-01 00:00:00     2,474.00
2012-06-01 01:00:00     2,349.00
2012-06-01 02:00:00     2,291.00
2012-06-01 03:00:00     2,281.00
2012-06-01 04:00:00     2,343.00
2012-06-01 05:00:00     2,518.00
2012-06-01 06:00:00     2,934.00
2012-06-01 07:00:00     3,235.00
2012-06-01 08:00:00     3,348.00
2012-06-01 09:00:00     3,405.00
2012-06-01 10:00:00     3,459.00
2012-06-01 11:00:00     3,479.00
2012-06-01 12:00:00     3,478.00
2012-06-01 13:00:00     3,495.00
2012-06-01 14:00:00     3,473.00
2012-06-01 15:00:00     3,439.00
2012-06-01 16:00:00     3,404.00
2012-06-01 17:00:00     3,337.00
2012-06-01 18:00:00     3,291.00
2012-06-01 19:00:00     3,261.00
2012-06-01 20:00:00     3,309.00
2012-06-01 21:00:00     3,197.00
2012-06-01 22:00:00     2,916.00
2012-06-01 23:00:00     2,619.00
```

The example below specifies a stop time that includes all of the times on the last day in our ts_data set:

```
ts_data_load['2012-1':'2012-2-28']
```

The output will be

```
load
2012-01-01 00:00:00      2,698.00
2012-01-01 01:00:00      2,558.00
2012-01-01 02:00:00      2,444.00
2012-01-01 03:00:00      2,402.00
2012-01-01 04:00:00      2,403.00
2012-01-01 05:00:00      2,453.00
2012-01-01 06:00:00      2,560.00
2012-01-01 07:00:00      2,719.00
2012-01-01 08:00:00      2,916.00
2012-01-01 09:00:00      3,105.00
2012-01-01 10:00:00      3,174.00
2012-01-01 11:00:00      3,180.00
2012-01-01 12:00:00      3,184.00
2012-01-01 13:00:00      3,147.00
2012-01-01 14:00:00      3,122.00
2012-01-01 15:00:00      3,137.00
2012-01-01 16:00:00      3,486.00
2012-01-01 17:00:00      3,717.00
2012-01-01 18:00:00      3,659.00
2012-01-01 19:00:00      3,513.00
2012-01-01 20:00:00      3,344.00
2012-01-01 21:00:00      3,129.00
2012-01-01 22:00:00      2,873.00
2012-01-01 23:00:00      2,639.00
2012-01-02 00:00:00      2,458.00
2012-01-02 01:00:00      2,354.00
2012-01-02 02:00:00      2,294.00
2012-01-02 03:00:00      2,288.00
2012-01-02 04:00:00      2,353.00
2012-01-02 05:00:00      2,503.00
...          ...
2012-02-27 18:00:00      3,966.00
2012-02-27 19:00:00      3,845.00
2012-02-27 20:00:00      3,626.00
2012-02-27 21:00:00      3,355.00
2012-02-27 22:00:00      3,070.00
2012-02-27 23:00:00      2,837.00
2012-02-28 00:00:00      2,681.00
2012-02-28 01:00:00      2,584.00
2012-02-28 02:00:00      2,539.00
2012-02-28 03:00:00      2,535.00
2012-02-28 04:00:00      2,626.00
2012-02-28 05:00:00      2,916.00
```

```
2012-02-28 06:00:00      3,316.00
2012-02-28 07:00:00      3,524.00
2012-02-28 08:00:00      3,594.00
2012-02-28 09:00:00      3,615.00
2012-02-28 10:00:00      3,600.00
2012-02-28 11:00:00      3,579.00
2012-02-28 12:00:00      3,506.00
2012-02-28 13:00:00      3,478.00
2012-02-28 14:00:00      3,429.00
2012-02-28 15:00:00      3,406.00
2012-02-28 16:00:00      3,477.00
2012-02-28 17:00:00      3,742.00
2012-02-28 18:00:00      3,927.00
2012-02-28 19:00:00      3,858.00
2012-02-28 20:00:00      3,687.00
2012-02-28 21:00:00      3,420.00
2012-02-28 22:00:00      3,122.00
2012-02-28 23:00:00      2,875.00
1416 rows × 1 columns
```

The following example specifies an exact stop time:

```
ts_data_load['2012-1':'2012-1-2 00:00:00']
```

The output will be

```
load
2012-01-01 00:00:00      2,698.00
2012-01-01 01:00:00      2,558.00
2012-01-01 02:00:00      2,444.00
2012-01-01 03:00:00      2,402.00
2012-01-01 04:00:00      2,403.00
2012-01-01 05:00:00      2,453.00
2012-01-01 06:00:00      2,560.00
2012-01-01 07:00:00      2,719.00
2012-01-01 08:00:00      2,916.00
2012-01-01 09:00:00      3,105.00
2012-01-01 10:00:00      3,174.00
2012-01-01 11:00:00      3,180.00
2012-01-01 12:00:00      3,184.00
2012-01-01 13:00:00      3,147.00
2012-01-01 14:00:00      3,122.00
2012-01-01 15:00:00      3,137.00
2012-01-01 16:00:00      3,486.00
2012-01-01 17:00:00      3,717.00
2012-01-01 18:00:00      3,659.00
2012-01-01 19:00:00      3,513.00
2012-01-01 20:00:00      3,344.00
2012-01-01 21:00:00      3,129.00
2012-01-01 22:00:00      2,873.00
2012-01-01 23:00:00      2,639.00
2012-01-02 00:00:00      2,458.00
```

There is also another function that you can use with time series data sets and that is similar to slicing: `truncate()`. The `truncate` function assumes a 0 value for any unspecified date component in a `DatetimeIndex`, in contrast to slicing, which returns any partially matching dates:

```
ts_data_load.truncate(before='2013-11-01', after='2013-11-02')
```

The output will be

```
                        load
2013-11-01 00:00:00     2,506.00
2013-11-01 01:00:00     2,419.00
2013-11-01 02:00:00     2,369.00
2013-11-01 03:00:00     2,349.00
2013-11-01 04:00:00     2,425.00
2013-11-01 05:00:00     2,671.00
2013-11-01 06:00:00     3,143.00
2013-11-01 07:00:00     3,438.00
2013-11-01 08:00:00     3,486.00
2013-11-01 09:00:00     3,541.00
2013-11-01 10:00:00     3,591.00
2013-11-01 11:00:00     3,585.00
2013-11-01 12:00:00     3,532.00
2013-11-01 13:00:00     3,491.00
2013-11-01 14:00:00     3,430.00
2013-11-01 15:00:00     3,358.00
2013-11-01 16:00:00     3,347.00
2013-11-01 17:00:00     3,478.00
2013-11-01 18:00:00     3,636.00
2013-11-01 19:00:00     3,501.00
2013-11-01 20:00:00     3,345.00
2013-11-01 21:00:00     3,131.00
2013-11-01 22:00:00     2,883.00
2013-11-01 23:00:00     2,626.00
2013-11-02 00:00:00     2,447.00
```

Time/Date Components

Finally, when dealing with time series data, it is important to remember all date and time properties that you can access from `Timestamp` or `DatetimeIndex`. Table 3.3 summarizes all of them for you.

Table 3.3: Date and time properties from `Timestamp` and `DatetimeIndex`

PROPERTY	DESCRIPTION
year	The year of the datetime
month	The month of the datetime
day	The days of the datetime
hour	The hour of the datetime
minute	The minutes of the datetime
second	The seconds of the datetime
microsecond	The microseconds of the datetime
nanosecond	The nanoseconds of the datetime
date	Returns `datetime.date` (does not contain timezone information)
time	Returns `datetime.time` (does not contain timezone information)
timetz	Returns `datetime.time` as local time with timezone information
dayofyear	The ordinal day of year
weekofyear	The week ordinal of the year
week	The week ordinal of the year
dayofweek	The number of the day of the week, with Monday = 0, Sunday = 6
weekday	The number of the day of the week with Monday = 0, Sunday = 6
quarter	Quarter of the date: Jan–Mar = 1, Apr–Jun = 2, etc.
days _ in _ month	The number of days in the month of the datetime
is _ month _ start	Logical indicating if first day of month (defined by frequency)
is _ month _ end	Logical indicating if last day of month (defined by frequency)
is _ quarter _ start	Logical indicating if first day of quarter (defined by frequency)
is _ quarter _ end	Logical indicating if last day of quarter (defined by frequency)
is _ year _ start	Logical indicating if first day of year (defined by frequency)
is _ year _ end	Logical indicating if last day of year (defined by frequency)
is _ leap _ year	Logical indicating if the date belongs to a leap year

Frequency Conversion

In the following examples, we use all the variables in the ts_data set (both load and temp variables) to understand how data scientists can apply a frequency conversion on their time series data.

First of all, let's load the ts_data set with and visualize the first 10 rows of both variables:

```
ts_data = load_data(data_dir)
ts_data.head(10)
```

The output will be

	Load	temp
2012-01-01 00:00:00	2,698.00	32.00
2012-01-01 01:00:00	2,558.00	32.67
2012-01-01 02:00:00	2,444.00	30.00
2012-01-01 03:00:00	2,402.00	31.00
2012-01-01 04:00:00	2,403.00	32.00
2012-01-01 05:00:00	2,453.00	31.33
2012-01-01 06:00:00	2,560.00	30.00
2012-01-01 07:00:00	2,719.00	29.00
2012-01-01 08:00:00	2,916.00	29.00
2012-01-01 09:00:00	3,105.00	33.33

The primary function for changing frequencies is the asfreq() method. This method converts a time series to specified frequency and optionally provides a filling method to pad/backfill missing values. For a DatetimeIndex, this is basically just a thin but convenient wrapper around reindex() that generates a date _ range and calls reindex, as shown in the following example:

```
daily_ts_data = ts_data.asfreq(pd.offsets.BDay())
daily_ts_data.head(5)
```

Running this example prints the first five rows of the transformed ts_data_ daily set:

	load	temp
2012-01-02	2,458.00	43.67
2012-01-03	2,780.00	26.33
2012-01-04	3,184.00	6.00
2012-01-05	3,014.00	22.33
2012-01-06	2,992.00	17.00

The asfreq function optionally provides a filling method to pad/backfill missing values. It returns the original data conformed to a new index with the specified frequency, as shown below:

```
daily_ts_data.asfreq(pd.offsets.BDay(), method='pad')
daily_ts_data.head(5)
```

Running this example prints the first five rows of the transformed `daily_ts_data` set with the 'pad' method:

```
                load        temp
2012-01-02      2,458.00    43.67
2012-01-03      2,780.00    26.33
2012-01-04      3,184.00    6.00
2012-01-05      3,014.00    22.33
2012-01-06      2,992.00    17.00
```

For further information on pandas Support for Time Series, visit the website `pandas.pydata.org/docs/user_guide/timeseries.html`.

Next, I will discuss how you can explore and analyze your time series data sets and how you can pre-process and perform feature engineering to enrich your time series data sets.

Time Series Exploration and Understanding

In the following sections, you will learn the first steps to take to explore, analyze, and understand time series data. We will focus on these topics:

- How to get started with time series data analysis
- How to calculate and review summary statistics for time series data
- How to perform data cleaning of missing periods in time series
- How to perform time series data normalization and standardization

How to Get Started with Time Series Data Analysis

As you saw in the first section of this chapter, pandas has proven very effective as a tool for handling and working with time series data: pandas has some in-built `datetime` functions that make it simple to work with time series data, because in this type of data set, time is the most important variable and dimension that data scientists can leverage to gain useful insights.

Now let's have a look at our ts_data set to see what kind of data we have by using the `head()` function. This function returns the first *n* rows for your time series data set, and it is useful for quickly getting an overview of the type of data and its structure in your data set:

```
ts_data.head(10)
```

As we see below, the data set has been classified into three columns: the hourly time stamp column, the load column, and the temperature column:

```
                        load        temp
2012-01-01 00:00:00     2,698.0     32.0
2012-01-01 01:00:00     2,558.0     32.7
```

Continues

(continued)

```
2012-01-01 02:00:00    2,444.0    30.0
2012-01-01 03:00:00    2,402.0    31.0
2012-01-01 04:00:00    2,403.0    32.0
2012-01-01 05:00:00    2,453.0    31.3
2012-01-01 06:00:00    2,560.0    30.0
2012-01-01 07:00:00    2,719.0    29.0
2012-01-01 08:00:00    2,916.0    29.0
2012-01-01 09:00:00    3,105.0    33.3
```

As a second step, it is important to get a summary of your time series data set in case there are rows with empty values. We can do this by using the `isna()` function in Python. This function takes a scalar or array-like objects and indicates whether values are missing, as the example below shows us:

```
ts_data.isna().sum()
```

The output will be

```
load    0
temp    0
dtype: int64
```

As we can see, we do not have empty values in our data set. The next step is to understand the format of our variables. pandas `.dtypes` method makes this possible, as it returns a series with the data type of each column:

```
ts_data.dtypes
```

The output will be

```
load    float64
temp    float64
dtype: object
```

We can see that both the load and temp columns are `float64`, which is a floating-point number that occupies 64 bits of storage. Floats represent real numbers and are written with a decimal point dividing the integer and fractional parts.

As the next steps, we need to calculate and review some summary statistics for our time series data set: summary statistics summarize large amounts of data by describing key characteristics such as the average, distribution, potential correlation, or dependence. Calculating descriptive statistics on time series can help give you an idea of the distribution and spread of values. This may help with ideas of data scaling and even data cleaning that we can perform later as part of preparing our data set for modeling.

In pandas, descriptive statistics include those useful indicators that summarize the central tendency, dispersion, and shape of a data set's distribution, excluding NaN values. The `describe()` function creates a summary of the loaded time series including mean, standard deviation, median, minimum, and maximum:

```
ts_data.describe()
```

The output will be

```
        load        temp
count   26,304.00   26,304.00
mean    3,303.77    47.77
std     564.58      19.34
min     1,979.00    -13.67
25%     2,867.00    32.67
50%     3,364.00    48.33
75%     3,690.00    63.67
max     5,224.00    95.00
```

For numeric data, the result index consists of count, mean, std, min, and max statistics as well as lower, 50, and upper percentiles. By default, the lower percentile is 25 and the upper percentile is 75. The 50 percentile is the same as the median.

On the other hand, for object data (e.g., strings or time stamps), the result index presents count, unique, top, and freq. Data scientists need to keep in mind that the top value does not represent the most common value; the freq represents the most common value. Additional important information are included time stamps, such as the first and last items. If multiple object values have the highest count, then the count and top results will be arbitrarily chosen from among those with the highest count.

Finally, for mixed data types presented in a DataFrame, the default option is to return only an evaluation of numeric columns. If the DataFrame contains only object and categorical data without any numeric columns, the default option is to show an overview of both the object and categorical columns. If include=all is given as an option, the final result will contain an overview of attributes of each type.

As you learned in Chapter 1, "Overview of Time Series Forecasting," generally time series often contain specific time-related information and features such as the following:

- *Trend*: This feature describes a visible change of time series values with respect to higher or lower values over a prolonged (but not seasonal or cyclical) period of time.

- *Seasonality*: This feature describes a recurring and persistent pattern in a time series within a fixed time period.

- *Cyclic*: This feature describes recurring and persistent patterns of upward and downward changes in your time series data, but it does not show a fixed pattern.

- *Noise*: This feature describes irregular values in your time series data, and it is also called noise, as it does not present a recurring and persistent pattern.

By using the `statsmodels` Python module, which has a `tsa` (time series analysis) package as well as the `seasonal_decompose()` function, we can visualize these for components and gain additional insights from our time series data.

The `statsmodels.tsa` package includes model classes and capabilities that are helpful to handle time series data. Some model examples are univariate autoregressive (AR) models, vector autoregressive (VAR) models, and univariate autoregressive moving average (ARMA) models. Some examples of non-linear models are Markov switching dynamic regression and autoregression.

The `statsmodels.tsa` package also contains descriptive statistics for time series, such as, autocorrelation, partial autocorrelation function, and periodogram. It also offers techniques to work with autoregressive and moving average lag-polynomials (`statsmodels.org/devel/tsa.html`).

Let's use the `statsmodels` and `seasonal_decompose()` functions to extract and visualize our ts_data set components only for the load variable, which is our target variable. First, let's import all necessary packages:

```
# import necessary Python packages
import statsmodels.api as sm
import warnings
import matplotlib
import matplotlib.pyplot as plt
import matplotlib.dates as mdates

%matplotlib inline

warnings.filterwarnings("ignore")
```

For practical reasons, we will visualize only a subset of our load data set:

```
ts_data_load = ts_data['load']
decomposition = sm.tsa.seasonal_decompose
(load['2012-07-01':'2012-12-31'], model = 'additive')

fig = decomposition.plot()
matplotlib.rcParams['figure.figsize'] = [10.0, 6.0]
```

The output will be the plot illustrated in Figure 3.2.

From Figure 3.2, we can observe a much clearer plot showing us that the seasonality is following a consistent pattern and the trend is following an irregular pattern. In order to investigate the trend in our time series a bit more, we can plot the trend alongside the observed time series.

To do this, we will use Matplotlib's `.YearLocator()`, which is a function that makes ticks on a given day of each year that is a multiple of base. For the plot, we set each year to begin from the month of January (month = 1) and month as the minor locator showing ticks for every three months (intervals = 3). Then we plot our data set using the index of the DataFrame as x-axis and the load variable for the y-axis. We performed the same steps for the trend observations.

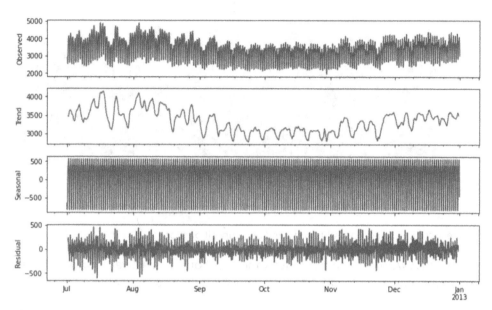

Figure 3.2: Time series decomposition plot for the load data set (time range: `2012-07-01 to 2012-12-31`)

```
decomposition = sm.tsa.seasonal_decompose(load, model = 'additive')

fig, ax = plt.subplots()
ax.grid(True)

year = mdates.YearLocator(month=1)
month = mdates.MonthLocator(interval=1)

year_format = mdates.DateFormatter('%Y')
month_format = mdates.DateFormatter('%m')

ax.xaxis.set_minor_locator(month)
ax.xaxis.grid(True, which = 'minor')
ax.xaxis.set_major_locator(year)
ax.xaxis.set_major_formatter(year_format)

plt.plot(load.index, load, c='blue')
plt.plot(decomposition.trend.index, decomposition.trend, c='white')
```

The output will be the plot illustrated in Figure 3.3.

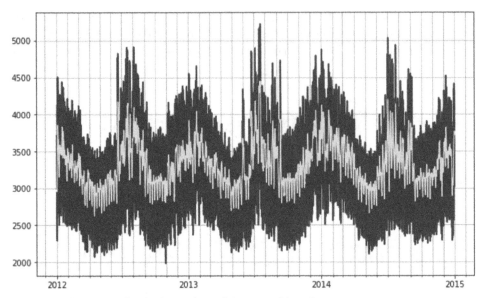

Figure 3.3: Time series load value and trend decomposition plot

Now that we have a better understanding of our data set, let's learn different techniques for data cleaning and interpolation for missing values.

Data Cleaning of Missing Values in the Time Series

Handling time series data involves cleaning, scaling, and even deep pre-processing operations before you obtain any meaningful results. In the following few paragraphs, we discuss some of the most common techniques used to clean, rescale, and prepare your time series data.

Missing values are represented in the time series by a sequence gap in the time stamp variable or in other values. Missing values may appear in time series for a number of different reasons, including recording failures or inaccurate recording due to a problem at the data source level or in the data pipeline.

To identify these time stamp gaps, first we create an index of time periods that we would expect to be in the time series. If we look at our ts_data set again, we can fill missing values by interpolating between the two closest non-missing values. The `dataframe.interpolate()` function is used to fill NA values in the DataFrame or series: it is a very effective function to fill the missing values. However, only `method = 'linear'` is supported for DataFrames and series with a MultiIndex.

In the example below, we use a quadratic function and set a limit of 8. This limit means that if more than 8 missing values occur consecutively, the missing values are not interpolated and they remain missing. This is to avoid spurious interpolation between very distant time periods. Moreover, we set `limit _`

direction ='both': limit direction for interpolation can be set as 'forward',
'backward', or 'both'; the default value is 'forward':

```
ts_data_load.interpolate(limit = 8, method ='linear', limit_direction
='both')
```

The output will be

```
2012-01-01 00:00:00    2,698.00
2012-01-01 01:00:00    2,558.00
2012-01-01 02:00:00    2,444.00
2012-01-01 03:00:00    2,402.00
2012-01-01 04:00:00    2,403.00
2012-01-01 05:00:00    2,453.00
2012-01-01 06:00:00    2,560.00
2012-01-01 07:00:00    2,719.00
2012-01-01 08:00:00    2,916.00
2012-01-01 09:00:00    3,105.00
2012-01-01 10:00:00    3,174.00
2012-01-01 11:00:00    3,180.00
2012-01-01 12:00:00    3,184.00
2012-01-01 13:00:00    3,147.00
2012-01-01 14:00:00    3,122.00
2012-01-01 15:00:00    3,137.00
2012-01-01 16:00:00    3,486.00
2012-01-01 17:00:00    3,717.00
2012-01-01 18:00:00    3,659.00
2012-01-01 19:00:00    3,513.00
2012-01-01 20:00:00    3,344.00
2012-01-01 21:00:00    3,129.00
2012-01-01 22:00:00    2,873.00
2012-01-01 23:00:00    2,639.00
2012-01-02 00:00:00    2,458.00
2012-01-02 01:00:00    2,354.00
2012-01-02 02:00:00    2,294.00
2012-01-02 03:00:00    2,288.00
2012-01-02 04:00:00    2,353.00
2012-01-02 05:00:00    2,503.00
                         . . .
2014-12-30 18:00:00    4,374.00
2014-12-30 19:00:00    4,270.00
2014-12-30 20:00:00    4,140.00
2014-12-30 21:00:00    3,895.00
2014-12-30 22:00:00    3,571.00
2014-12-30 23:00:00    3,313.00
2014-12-31 00:00:00    3,149.00
2014-12-31 01:00:00    3,055.00
2014-12-31 02:00:00    3,014.00
2014-12-31 03:00:00    3,025.00
2014-12-31 04:00:00    3,115.00
2014-12-31 05:00:00    3,337.00
```

Continues

(continued)

```
2014-12-31 06:00:00    3,660.00
2014-12-31 07:00:00    3,906.00
2014-12-31 08:00:00    4,043.00
2014-12-31 09:00:00    4,077.00
2014-12-31 10:00:00    4,073.00
2014-12-31 11:00:00    4,030.00
2014-12-31 12:00:00    3,982.00
2014-12-31 13:00:00    3,933.00
2014-12-31 14:00:00    3,893.00
2014-12-31 15:00:00    3,912.00
2014-12-31 16:00:00    4,141.00
2014-12-31 17:00:00    4,319.00
2014-12-31 18:00:00    4,199.00
2014-12-31 19:00:00    4,012.00
2014-12-31 20:00:00    3,856.00
2014-12-31 21:00:00    3,671.00
2014-12-31 22:00:00    3,499.00
2014-12-31 23:00:00    3,345.00
Freq: H, Name: load, Length: 26304, dtype: float64
```

In the next example, we fill missing temperature values with a common value of 0 by leveraging the `scipy.stats` package. Many useful statistics functions are located in this sub-package `scipy.stats` (you can see a complete list of this functions using `info(stats)` and you can find additional information and tutorials at `docs.scipy.org/doc/scipy/reference/tutorial/stats.html`):

```
from scipy import stats
temp_mode = np.asscalar(stats.mode(ts_data['temp']).mode)
ts_data['temp'] = ts_data['temp'].fillna(temp_mode)
ts_data.isnull().sum()
```

The output will be

```
load    0
temp    0
dtype: int64
```

As you can see from your results, the number of missing values has now been reduced, if not completely eliminated. If you still have records containing remaining missing values in your data set, you can remove them later after model features have been created. Now let's see how we can normalize time series data in the next section.

Time Series Data Normalization and Standardization

Normalization is the process of rescaling data from the original scale so that all values are within the range of 0 and 1, and it involves the estimation of the minimum and maximum available values in your data set. It can be useful, and even required in some machine learning algorithms, when your time series data has input values and features with differing measurements and dimensions.

For machine learning algorithms, such as k-nearest neighbors, which uses distance estimates, linear regression and neural networks that process a weight calibration on input values, normalization is necessary. However, it is important to keep in mind that if your time series presents a visible trend, estimating these expected values may be difficult and normalization may not be the best method to use on your problem.

You can normalize your data set using the scikit-learn object MinMaxScaler: `sklearn.preprocessing.MinMaxScaler`. This estimator transforms each feature individually such that it is in the given range, like the range of 0 and 1 (`www.scikit-learn.org/stable/modules/generated/sklearn.preprocessing.MinMaxScaler.html`).

This transformation can be also inverted, as sometimes it can be useful for data scientists to convert predictions back into their original scale for reporting or plotting. This can be done by calling the `inverse _ transform()` function. Below is an example of normalizing the ts_data set: as the example shows, the scaler requires data to be provided as a matrix of rows and columns. The load data is loaded as a pandas DataFrame. It must then be reshaped into a matrix of one column:

```
from pandas import Series
from sklearn.preprocessing import MinMaxScaler

# prepare data for normalization
values = load.values
values = values.reshape((len(values), 1))

# train the normalization
scaler = MinMaxScaler(feature_range=(0, 1))
scaler = scaler.fit(values)
print('Min: %f, Max: %f' % (scaler.data_min_, scaler.data_max_))
```

The output will be

```
Min: 1979.000000, Max: 5224.000000
```

The reshaped data set is then used to fit the scaler, the data set is normalized, then the normalization transform is inverted to show the original values again:

```
# normalize the data set and print the first 5 rows
normalized = scaler.transform(values)
for i in range(5):
    print(normalized[i])

# inverse transform and print the first 5 rows
inversed = scaler.inverse_transform(normalized)
```

Continues

(*continued*)
```
for i in range(5):
    print(inversed[i])
```

The output will be

```
[0.22]
[0.18]
[0.14]
[0.13]
[0.13]
[2698.]
[2558.]
[2444.]
[2402.]
[2403.]
```

Running the example prints the first five rows from the loaded data set, shows the same five values in their normalized form, then the values back in their original scale using the inverse transform capability. There is another type of rescaling that is more robust to new values being outside the range of expected values: standardization.

Standardizing a data set involves rescaling the distribution of values so that the mean of observed values is 0 and the standard deviation is 1. This process implies subtracting the mean value or centering the data. Like normalization, standardization can be useful, and even required in some machine learning algorithms when your time series data has input values with differing dimensions. Standardization assumes that your observations fit a Gaussian distribution (bell curve) with a well-behaved mean and standard deviation. This includes algorithms like support vector machines and linear and logistic regression and other algorithms that assume or have improved performance with Gaussian data.

In order to apply a standardization process to a data set, data scientists need to accurately estimate the mean and standard deviation of the values in their data. You can standardize your data set using the scikit-learn object StandardScaler: `sklearn.preprocessing.StandardScaler`. This capability standardizes features by removing the mean and scaling to unit variance. Centering and scaling happen independently on each feature by computing the relevant statistics on the samples in the training set. Mean and standard deviation are then stored to be used on later data using `transform` (scikit-learn.org/stable/modules/generated/sklearn.preprocessing.StandardScaler.html).

Below is an example of standardizing our load data set:

```
# Standardize time series data
from sklearn.preprocessing import StandardScaler
from math import sqrt
```

```
# prepare data for standardization
values = load.values
values = values.reshape((len(values), 1))

# train the standardization
scaler = StandardScaler()
scaler = scaler.fit(values)
print('Mean: %f, StandardDeviation: %f' % (scaler.mean_, sqrt(scaler.
var_)))
```

The output will be

```
Mean: 3303.769199, StandardDeviation: 564.568521
```

```
# standardization the data set and print the first 5 rows
normalized = scaler.transform(values)
for i in range(5):
    print(normalized[i])

# inverse transform and print the first 5 rows
inversed = scaler.inverse_transform(normalized)
for i in range(5):
        print(inversed[i])
```

The output will be

```
[-1.07]
[-1.32]
[-1.52]
[-1.6]
[-1.6]
[2698.]
[2558.]
[2444.]
[2402.]
[2403.]
```

Running the example prints the first five rows of the data set, prints the same values standardized, then prints the values back in their original scale.

In the next section, you will discover how to perform feature engineering on time series data with Python to model your time series problem with machine learning algorithms.

Time Series Feature Engineering

As mentioned in Chapter 1, time series data needs to be remodeled as a supervised learning data set before we can start using machine learning algorithms. In the field of artificial intelligence and machine learning, supervised learning is defined as a type of method in which data scientists need to provide both input and output data to their machine learning algorithms.

In the next step, the supervised learning algorithm (for example, linear regression for regression problems, random forest for classification and regression problems, support vector machines for classification problems) examines the training data set and produces a mathematical function, which can be used for inferring new examples and predictions.

In time series, data scientists have to construct the output of their model by identifying the variable that they need to predict at a future date (e.g., future number of sales next Monday) and then leverage historical data and feature engineering to create input variables that will be used to make predictions for that future date.

Feature engineering efforts mainly have two goals:

- *Creating the correct input data set to feed the machine learning algorithm*: In this case, the purpose of feature engineering in time series forecasting is to create input features from historical and row data and shape the data set as a supervised learning problem.

- *Increasing the performance of machine learning models*: The second most important goal of feature engineering is about generating valid relationships between input features and the output feature or target variable to be predicted. In this way, the performance of machine learning models can be improved.

We will cover four different categories of time features that are extremely helpful in time series scenarios:

- Date time features
- Lag features and window features
- Rolling window statistics
- Expanding window statistics

In the next few sections, we will discuss each of these time features in more detail, and I will explain them with real-word examples in Python.

Date Time Features

Date Time Features are features created from the time stamp value of each observation. A few examples of these features are the integer hour, month, day of week for each observation. Data scientists can perform these date time features transformations using pandas and adding new columns (hour, month, and day of week columns) to their original data set where hour, month, and day of week information is extracted from the time stamp value for each observation.

Below is some sample Python code to perform this with our ts_data set:

```
ts_data['hour'] = [ts_data.index[i].hour for i in range(len(ts_data))]
ts_data['month'] = [ts_data.index[i].month for i in range(len(ts_data))]
ts_data['dayofweek'] = [ts_data.index[i].day for i in range(len(ts_
data))]
print(ts_data.head(5))
```

Running this example prints the first five rows of the transformed data set:

```
                       load     temp     hour    month    dayofweek
2012-01-01 00:00:00  2,698.00  32.00     0       1        1
2012-01-01 01:00:00  2,558.00  32.67     1       1        1
2012-01-01 02:00:00  2,444.00  30.00     2       1        1
2012-01-01 03:00:00  2,402.00  31.00     3       1        1
2012-01-01 04:00:00  2,403.00  32.00     4       1        1
```

By leveraging this additional knowledge, such as hour, month, and day of week values, data scientists can gain additional insights on their data and on the relationship between the input features and the output feature and eventually build a better model for their time series forecasting solutions. Here are some other examples of features that can be built and generate additional and important information:

- Weekend or not
- Minutes in a day
- Daylight savings or not
- Public holiday or not
- Quarter of the year
- Hour of day
- Before or after business hours
- Season of the year

As you can observe from the examples above, date time features are not limited only to integer values. Data scientists can also build binary features, such as a feature in which, if the time stamp information is before business hours, its value equals 1; if the time stamp information is after business hours, its value equals 0. Finally, when dealing with time series data, it is important to remember all date and time properties that you can access from Timestamp or DatetimeIndex. (Table 3.2 in this chapter summarizes all of them for you).

Date time features represent a very useful way for data scientists to start their feature engineering work with time series data. In the next section, I will introduce an additional approach to build input features for your data set: lag and window features. In order to build these features, data scientists need to leverage and extract the values of a series in previous or future periods.

Lag Features and Window Features

Lag features are values at prior timesteps that are considered useful because they are created on the assumption that what happened in the past can influence or contain a sort of intrinsic information about the future. For example, it can be beneficial to generate features for sales that happened in previous days at 4:00 p.m. if you want to predict similar sales at 4:00 p.m. the day after.

An interesting category of lag features is called *nested* lag features: in order to create nested lag features, data scientists need to identify a fixed time period in the past and group feature values by that time period—for example, the number of items sold in the previous two hours, previous three days, and previous week.

The pandas library provides the `shift()` function to help create these shifted or lag features from a time series data set: this function shifts an index by desired number of periods with an optional time frequency. The `shift` method accepts a `freq` argument which can accept a `DateOffset` class or a `timedelta`-like object or also an `offset alias`. Offset alias is an important concept that data scientists can leverage when they deal with time series data because it represents the number of string aliases that are given to useful common time series frequencies, as summarized in Table 3.4.

Table 3.4: Offset aliases supported in Python

OFFSET ALIAS	DESCRIPTION
B	Business day frequency
C	Custom business day frequency
D	Calendar day frequency
W	Weekly frequency
M	Month end frequency
SM	Semi-month end frequency (15th and end of month)
BM	Business month end frequency
CBM	Custom business month end frequency
MS	Month start frequency
SMS	Semi-month start frequency (1st and 15th)
BMS	Business month start frequency
CBMS	Custom business month start frequency
Q	Quarter end frequency
BQ	Business quarter end frequency

OFFSET ALIAS	DESCRIPTION
QS	Quarter start frequency
BQS	Business quarter start frequency
A, Y	Year-end frequency
BA, BY	Business year end frequency
AS, YS	Year start frequency
BAS, BYS	Business years start frequency
BH	Business hour frequency
H	Hourly frequency
T, min	Minutely frequency
S	Secondly frequency
L, ms	Milliseconds
U, us	Microseconds
N	Nanoseconds

Below is an example of how to use offset aliases on the ts_data set:

```
ts_data_shift = ts_data.shift(4, freq=pd.offsets.BDay())
ts_data_shift.head(5)
```

Running this example prints the first five rows of the transformed ts_data_shift data set:

```
                       load      temp
2012-01-05 00:00:00    2,698.00  32.00
2012-01-05 01:00:00    2,558.00  32.67
2012-01-05 02:00:00    2,444.00  30.00
2012-01-05 03:00:00    2,402.00  31.00
2012-01-05 04:00:00    2,403.00  32.00
```

Rather than changing the alignment of the data and the index, `DataFrame` and `Series` objects also have a `tshift()` convenience method that changes all the dates in the index by a specified number of offsets, using the index's frequency if available, as shown in the following example:

```
ts_data_shift_2 = ts_data.tshift(6, freq='D')
ts_data_shift_2.head(5)
```

Running this example prints the first five rows of the transformed ts_data_shift_2 data set:

```
                       load      temp
2012-01-07 00:00:00    2,698.00  32.00
```

Continues

(continued)

```
2012-01-07 01:00:00     2,558.00    32.67
2012-01-07 02:00:00     2,444.00    30.00
2012-01-07 03:00:00     2,402.00    31.00
2012-01-07 04:00:00     2,403.00    32.00
```

Note that with `tshift`, the leading entry is a NaN value because the data is not being realigned. When the value for the frequency is not passed, then it shifts the index without realigning the data. If a frequency value is passed (in this case, the index must be date or datetime or it will raise a `NotImplementedError`), the index will be increased using the periods and the frequency value.

Shifting the data set by 1 creates a new "t" column and adds a NaN (unknown) value for the first row. The time series data set without a shift represents the "t+1". Let's make this concrete with an example. We can use our ts_data set to compute lagged load features, as follows:

```
def generated_lagged_features(ts_data, var, max_lag):
    for t in range(1, max_lag+1):
        ts_data[var+'_lag'+str(t)] = ts_data[var].shift(t, freq='1H')
```

In the sample code above, we first create a function called `generated_lagged_features` and then we generate eight additional lag features with one-hour frequency, as follows:

```
generated_lagged_features(ts_data, 'load', 8)
generated_lagged_features(ts_data, 'temp', 8)
print(ts_data.head(5))
```

Running this example prints the first five rows of the transformed data set:

```
                      load   temp   hour   month   dayofweek   load_lag1   \
2012-01-01 00:00:00  2,698.00 32.00     0      1         1          1        nan
2012-01-01 01:00:00  2,558.00 32.67     1      1         1          1     2,698.00
2012-01-01 02:00:00  2,444.00 30.00     2      1         1          1     2,558.00
2012-01-01 03:00:00  2,402.00 31.00     3      1         1          1     2,444.00
2012-01-01 04:00:00  2,403.00 32.00     4      1         1          1     2,402.00


                      load_lag2   load_lag3   load_lag4   load_lag5    ...
\
2012-01-01 00:00:00        nan         nan         nan         nan     ...
2012-01-01 01:00:00        nan         nan         nan         nan     ...
2012-01-01 02:00:00     2,698.00       nan         nan         nan     ...
2012-01-01 03:00:00     2,558.00    2,698.00       nan         nan     ...
2012-01-01 04:00:00     2,444.00    2,558.00    2,698.00       nan     ...


                      temp_lag1   temp_lag2   temp_lag3   temp_lag4   temp_
lag5   \
2012-01-01 00:00:00        nan         nan         nan         nan
nan
```

2012-01-01 01:00:00	32.00	nan	nan	nan
nan				
2012-01-01 02:00:00	32.67	32.00	nan	nan
nan				
2012-01-01 03:00:00	30.00	32.67	32.00	nan
nan				
2012-01-01 04:00:00	31.00	30.00	32.67	32.00
nan				

	temp_lag6	load_lag7	load_lag8	temp_lag7	temp_
lag8					
2012-01-01 00:00:00	nan	nan	nan	nan	
nan					
2012-01-01 01:00:00	nan	nan	nan	nan	
nan					
2012-01-01 02:00:00	nan	nan	nan	nan	
nan					
2012-01-01 03:00:00	nan	nan	nan	nan	
nan					
2012-01-01 04:00:00	nan	nan	nan	nan	
nan					

```
[5 rows x 21 columns]
```

The operation of adding lag features is called the sliding window method or Window Features: the above example shows how to apply a sliding window method with a window width of eight. Window Features are a summary of values over a fixed window of prior timesteps.

Depending on your time series scenario, you can expand the window width and include more lagged features. A common question that data scientists ask before performing the operation of adding lag features is how large to make the window. A good approach would be to build a series of different window widths and alternatively add and remove them from the data set to see which one has a more evident positive effect on your model performance.

Understanding the sliding method is very helpful to building an additional feature method called Rolling Window Statistics, which we will discuss in the next section.

Rolling Window Statistics

The main goal of building and using Rolling Window Statistics in a time series data set is to compute statistics on the values from a given data sample by defining a range that includes the sample itself as well as some specified number of samples before and after the sample used.

A crucial step when data scientists need to compute rolling statistics is to define a rolling window of observations: at each time point, data scientists need to obtain the observations in the rolling window and use them to compute the statistic they have decided to use. In the second step, they need to move on to the next time point and repeat the same computation on the next window of observations.

One of the more popular rolling statistics is the moving average. This takes a moving window of time and calculates the average or the mean of that time period as the current value. pandas provides a `rolling()` function that provides rolling window calculations, and it creates a new data structure with the window of values at each timestep. We can then perform statistical functions on the window of values collected for each timestep, such as calculating the mean.

Data scientists can use the `concat()` function in pandas to construct a new data set with just our new columns. This function concatenates pandas objects along a particular axis with an optional set logic along the other axes. It can also add a layer of hierarchical indexing on the concatenation axis, which may be useful if the labels are the same (or overlapping) on the passed axis number.

The example below demonstrates how to do this with pandas with a window size of 6:

```python
# create a rolling mean feature
from pandas import concat

load_val = ts_data[['load']]
shifted = load_val.shift(1)

window = shifted.rolling(window=6)
means = window.mean()
new_dataframe = concat([means, load_val], axis=1)
new_dataframe.columns = ['load_rol_mean', 'load']

print(new_dataframe.head(10))
```

Running the example above prints the first 10 rows of our new data set:

```
                     load_rol_mean load
2012-01-01 00:00:00            nan 2,698.00
2012-01-01 01:00:00            nan 2,558.00
2012-01-01 02:00:00            nan 2,444.00
2012-01-01 03:00:00            nan 2,402.00
2012-01-01 04:00:00            nan 2,403.00
2012-01-01 05:00:00            nan 2,453.00
2012-01-01 06:00:00       2,493.00 2,560.00
2012-01-01 07:00:00       2,470.00 2,719.00
2012-01-01 08:00:00       2,496.83 2,916.00
2012-01-01 09:00:00       2,575.50 3,105.00
```

Below is another example that shows how to create a window width of 4, use the `rolling()` function, and build a data set containing more summary statistics, such as the minimum, mean, and maximum value in the window:

```
# create rolling statistics features
from pandas import concat

load_val = ts_data[['load']]
width = 4
shifted = load_val.shift(width - 1)
window = shifted.rolling(window=width)

new_dataframe = pd.concat([window.min(),
window.mean(), window.max(), load_val], axis=1)
new_dataframe.columns = ['min', 'mean', 'max', 'load']

print(new_dataframe.head(10))
```

Running the code prints the first 10 rows of the new data set, with the new features just created, min, mean and max:

	min	mean	max	load
2012-01-01 00:00:00	nan	nan	nan	2,698.00
2012-01-01 01:00:00	nan	nan	nan	2,558.00
2012-01-01 02:00:00	nan	nan	nan	2,444.00
2012-01-01 03:00:00	nan	nan	nan	2,402.00
2012-01-01 04:00:00	nan	nan	nan	2,403.00
2012-01-01 05:00:00	nan	nan	nan	2,453.00
2012-01-01 06:00:00	2,402.00	2,525.50	2,698.00	2,560.00
2012-01-01 07:00:00	2,402.00	2,451.75	2,558.00	2,719.00
2012-01-01 08:00:00	2,402.00	2,425.50	2,453.00	2,916.00
2012-01-01 09:00:00	2,402.00	2,454.50	2,560.00	3,105.00

Another type of window feature that may be useful in time series forecasting scenarios is the Expanding Window Statistics feature: it includes all previous data in the series. We will discuss and learn how to build it in the next section.

Expanding Window Statistics

Expanding Window are features that include all previous data. pandas offers an `expanding()` function that provides expanding transformations and assembles sets of all prior values for each timestep: Python offers the same interface and capabilities for the `Rolling()` and `expanding()` functions.

Below is an example of calculating the minimum, mean, and maximum values of the expanding window on the ts_data set:

```
# create expanding window features
from pandas import concat
```

Continues

(continued)

```
load_val = ts_data[['load']]
window = load_val.expanding()
new_dataframe = concat([window.min(),
window.mean(), window.max(), load_val. shift(-1)], axis=1)
new_dataframe.columns = ['min', 'mean', 'max', 'load+1']
print(new_dataframe.head(10))
```

Running the example prints the first 10 rows of the new data set with the additional expanding window features:

```
                       min       mean      max      load+1
2012-01-01 00:00:00  2,698.00  2,698.00  2,698.00  2,558.00
2012-01-01 01:00:00  2,558.00  2,628.00  2,698.00  2,444.00
2012-01-01 02:00:00  2,444.00  2,566.67  2,698.00  2,402.00
2012-01-01 03:00:00  2,402.00  2,525.50  2,698.00  2,403.00
2012-01-01 04:00:00  2,402.00  2,501.00  2,698.00  2,453.00
2012-01-01 05:00:00  2,402.00  2,493.00  2,698.00  2,560.00
2012-01-01 06:00:00  2,402.00  2,502.57  2,698.00  2,719.00
2012-01-01 07:00:00  2,402.00  2,529.62  2,719.00  2,916.00
2012-01-01 08:00:00  2,402.00  2,572.56  2,916.00  3,105.00
2012-01-01 09:00:00  2,402.00  2,625.80  3,105.00  3,174.00
```

In this section, you have discovered how to use feature engineering to transform a time series data set into a supervised learning data set for machine learning and improve the performance of your machine learning models.

Conclusion

In this chapter, you learned the most important and basic steps to prepare your time series data for forecasting models. Good time series data preparation produces clean and well-curated data, which results in more accurate predictions. Specifically, in this chapter you learned the following:

- *Python for Time Series Data* – Python is a very powerful programming language to handle data, offering an assorted suite of libraries for time series data and excellent support for time series analysis. In this section, you learned how libraries such as SciPy, NumPy, Matplotlib, pandas, statsmodels, and scikit-learn can help you prepare, explore, and analyze your time series data.

- *Time Series Exploration and Understanding* – In this second section, you learned the first steps to take to explore, analyze, and understand time series data, such as how to calculate and review summary statistics for time series data, how to perform data cleaning of a missing period in time series, and how to perform time series data normalization and standardization.

- *Time Series Feature Engineering* – Feature engineering is the process of using historical row data to create additional variables and features for the final data set that you will use to train your model. In this final section of Chapter 3, you learned how to perform feature engineering on time series data.

In the next chapter, you will discover a suite of classical methods for time series forecasting that you can test on your forecasting problems. I will walk you through different methods such as autoregression, Autoregressive Moving Average, Autoregressive Integrated Moving Average, and Automated Machine Learning.

4

Introduction to Autoregressive and Automated Methods for Time Series Forecasting

Building forecasts is an integral part of any business, whether it is about revenue, inventory, online sales, or customer demand forecasting. Time series forecasting remains so fundamental because there are several problems and related data in the real world that present a time dimension.

Applying machine learning models to accelerate forecasts enables scalability, performance, and accuracy of intelligent solutions that can improve business operations. However, building machine learning models is often time consuming and complex with many factors to consider, such as iterating through algorithms, tuning machine learning hyperparameters, and applying feature engineering techniques. These options multiply with time series data as data scientists need to consider additional factors, such as trends, seasonality, holidays, and external economic variables.

In this chapter, you will discover a suite of classical methods for time series forecasting that you can test on your forecasting problems. The following paragraphs are structured to give you just enough information on each method to get started with a working code example and where to look to get more information on the method.

Specifically, we will look at the following classical methods:

- *Autoregression* – This time series technique assumes that future observations at next time stamp are related to the observations at prior time stamps through a linear relationship.

- *Moving Average* – This time series technique leverages previous forecast errors in a regression approach to forecast future observations at next time stamps. Together with the autoregression technique, the moving average approach is a crucial element of the more general autoregressive moving average and autoregressive integrated moving average models, which present a more sophisticated stochastic configuration.

- *Autoregressive Moving Average* – This time series technique assumes that future observations at next time stamp can be represented as a linear function of the observations and residual errors at prior time stamps.

- *Autoregressive Integrated Moving Average* – This time series technique assumes that future observations at next time stamp can be represented as a linear function of the differenced observations and residual errors at prior time stamps. We will also look at an extension of the autoregressive integrated moving average approach, called seasonal autoregressive integrated moving average with exogenous regressors.

- *Automated Machine Learning (Automated ML)* – This time series technique iterates through a portfolio of different machine learning algorithms for time series forecasting, while performing best model selection, hyperparameters tuning, and feature engineering for your scenario.

As you can see, classical time series forecasting methods are generally focused on the linear relationship between historical data and future results. Nevertheless, they perform well on a wide range of time series problems. Before exploring more sophisticated deep learning methods for time series (Chapter 5, "Introduction to Neural Networks for Time Series Forecasting"), it is a good idea to ensure you have exhausted classical linear time series forecasting methods.

Autoregression

Autoregression is a time series forecasting approach that depends only on the previous outputs of a time series: this technique assumes that future observations at next time stamp are related to the observations at prior time stamps through a linear relationship. In other words, in an autoregression, a value from a time series is regressed on previous values from that same time series. In this chapter of the book, we will learn how to implement an autoregression for time series forecasting with Python.

In an autoregression, the output value in the previous time stamp becomes the input value to predict the next time stamp value, and the errors follow the usual assumptions about errors in a simple linear regression model. In autoregressions, the number of preceding input values in the time series that are used to predict next time stamp value is called *order* (often we refer to the order with the letter *p*). This order value determines how many previous data points will be used: usually, data scientists estimate the p value by testing different values and observing which model resulted with the minimum Akaike information criterion (AIC). We will discuss both AIC and Bayesian information criterion (BIC) penalized-likelihood criteria later in this chapter.

Data scientists refer to autoregressions in which the current predicted value (output) is based on the immediately preceding value (input) as first order autoregression, as illustrated in Figure 4.1.

Sensor ID	Time Stamp	Value X	Value y
Sensor_1	01/01/2020	NaN	236
Sensor_1	01/01/2020	236	133
Sensor_1	01/02/2020	133	148
Sensor_1	01/03/2020	148	152
Sensor_1	01/04/2020	152	241
Sensor_1	01/05/2020	241 ← ?	Value to be regressed on previous value from that same time series

Figure 4.1: First order autoregression approach

If you need to predict the next time stamp value using the previous two values instead, then the approach is called a second order autoregression, as the next time stamp value will be predicted using as input two previous values, as illustrated in Figure 4.2.

Sensor ID	Time Stamp	Value X	Value y
Sensor_1	01/01/2020	NaN	236
Sensor_1	01/01/2020	236	133
Sensor_1	01/02/2020	133	148
Sensor_1	01/03/2020	148	152
Sensor_1	01/04/2020	152	241
Sensor_1	01/05/2020	241 ← ?	Value to be regressed on previous two value from that same time series

Figure 4.2: Second order autoregression approach

More generally, an *nth*-order autoregression is a multiple linear regression in which the value of the series at any time *t* is a linear function of the previous values in that same time series. Because of this serial dependence, another important aspect of autoregressions is *autocorrelation*: autocorrelation is a statistical property that occurs when a time series is linearly related to a previous or lagged version of itself.

Autocorrelation is a crucial concept for autoregressions as the stronger the correlation between the output (that is, the target variable that we need to predict) and a specific lagged variable (that is, a group of values at a prior time stamp used as input), the more weight that autoregression can put on that specific variable. So that variable is considered to have a strong predictive power.

Moreover, some regression methods, such as linear regression and ordinary least squares regression, rely on the implicit assumption that there is no presence of autocorrelation in the training data set used to feed the model. These methods, like linear regression and ordinary least squares regression, are defined as *parametric* methodologies, as the set of data used with them presents a normal distribution and their regression function is defined in terms of a finite number of unknown parameters that are estimated from the data.

For all these reasons, autocorrelation can help data scientists select the most appropriate method for their time series forecasting solutions. Furthermore, autocorrelation can be very useful to gain additional insights from your data and between your variables, and identify hidden patterns, such as seasonality and trend in time series data.

In order to check whether there is a presence of autocorrelation in your time series data, you can use two different built-in plots provided by pandas, called `lag_plot` (pandas.pydata.org/docs/reference/api/pandas.plotting.lag_plot .html) and `autocorrelation_plot` (pandas.pydata.org/docs/reference/api/ pandas.plotting.autocorrelation_plot.html).

These functions can be imported from `pandas.plotting` and take a Series or DataFrame as an argument. Both plots are visual checks that you can leverage to see if there is autocorrelation in your time series data set, as illustrated in Table 4.1.

Table 4.1: pandas.plotting.lag_plot API reference and description

API REFERENCE	pandas.plotting.lag_plot
Parameters	series : *time series*
	lag : *lag of the scatter plot, default 1*
	ax : *matplotlib axis object, optional*
	kwds : *matplotlib scatter method keyword arguments, optional*
Returns	class: *matplotlib.axis.Axes*

Below is an example of creating a lag plot for our ts_data_load set. First of all, let's import all necessary libraries and load our data into a pandas DataFrame:

```
# Import necessary libraries
import datetime as dt
import os
import warnings
from collections import UserDict
from glob import glob

import matplotlib.pyplot as plt
import numpy as np
import pandas as pd
from common.utils import load_data, mape
from IPython.display import Image

%matplotlib inline

pd.options.display.float_format = "{:,.2f}".format
np.set_printoptions(precision=2)
warnings.filterwarnings("ignore")

# Load the data from csv into a pandas dataframe
ts_data_load = load_data(data_dir)[['load']]
ts_data_load.head()
```

Now, let's create a lag plot for our ts_data_load set, by selecting only the load column:

```
# Import lag_plot function
from pandas.plotting import lag_plot
plt.figure()

# Pass the lag argument and plot the values.
# When lag=1 the plot is essentially data[:-1] vs. data[1:]
# Plot our ts_data_load set
lag_plot(ts_data_load)
```

Lag plots are used to check if a data set or time series is random: random data should not exhibit any structure in the lag plot. Let's look at the results illustrated in Figure 4.3.

In Figure 4.3, we can see a large concentration of energy values along a diagonal line of the plot. It clearly shows a relationship or some correlation between those observations of the data set. Moreover, we can use the autocorrelation _ plot, as described in Table 4.2.

Table 4.2: pandas.plotting.lag_plot API reference and description

API REFERENCE	pandas.plotting.autocorrelation_plot
Parameters	series : *time series*
	lag : *lag of the scatter plot, default 1*
	ax : *matplotlib axis object, optional*
	kwds : *keywords*
	Options to pass to matplotlib plotting method
Returns	class: *matplotlib.axis.Axes*

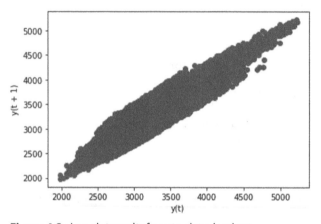

Figure 4.3: Lag plot results from ts_data_load set

Autocorrelation plots are also often applied by data scientists to check randomness in time series by computing autocorrelations for data values at fluctuating time lags. If time series is random, autocorrelation values should be near zero for all time lags. If time series is nonrandom, then one or more of the autocorrelations will be significantly non-zero.

Below is an example of creating an autocorrelation plot for our ts_data_load set.

```
# Import autocorrelation_plot function
from pandas.plotting import autocorrelation_plot
plt.figure()

# Pass the autocorrelation argument and plot the values
autocorrelation_plot(ts_data_load)
```

Let's look at the results illustrated in Figure 4.4.

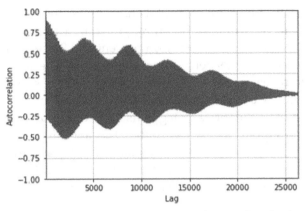

Figure 4.4: Autocorrelation plot results from ts_data_load set

Since our ts_data_load set is very granular and contains a large amount of hourly data points, we are not able to see the horizontal lines that are supposed to be displayed in the autocorrelation plot. For this reason, we can create a subset of our data set (for example, we can select the first week of August 2014) and apply again the autocorrelation plot function, as shown the sample code below:

```
# Create subset
ts_data_load_subset = ts_data_load['2014-08-01':'2014-08-07']

# Import autocorrelation _plot function
from pandas.plotting import autocorrelation_plot
plt.figure()

# Pass the autocorrelation argument and plot the values
autocorrelation_plot(ts_data_load_subset)
```

We can now look at the results of the autocorrelation plot, as illustrated in Figure 4.5.

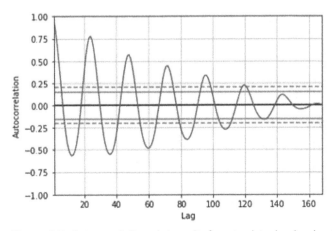

Figure 4.5: Autocorrelation plot results from ts_data_load_subset

As illustrated in Figure 4.5, the autocorrelation plot shows the value of the autocorrelation function on the vertical axis. It can range from –1 to 1. The horizontal lines displayed in the plot correspond to 95 percent and 99 percent confidence bands, and the dashed line is 99 percent confidence band. The autocorrelation plot is intended to reveal whether the data points of a time series are positively correlated, negatively correlated, or independent of each other.

A plot of the autocorrelation of a time series by lag is also called the autocorrelation function (ACF). Python support the ACF with the `plot_acf()` function from the statsmodels library (statsmodels.org/devel/generated/statsmodels.graphics.tsaplots.plot_acf.html). Below is an example of calculating and plotting the autocorrelation plot for our ts_data_load set using the `plot_acf()` function from the statsmodels library:

```
# Import plot_acf() function
from statsmodels.graphics.tsaplots import plot_acf

# Plot the acf function on the ts_data_load set
plot_acf(ts_data_load)
pyplot.show()
```

Let's run the same `plot_acf()` function on our ts_data_load_subset:

```
# Import plot_acf() function
from statsmodels.graphics.tsaplots import plot_acf

# Plot the acf function on the ts_data_load_subset
plot_acf(ts_data_load_subset)
pyplot.show()
```

Running these examples creates two 2D plots showing the lag value along the x-axis and the correlation on the y-axis between -1 and 1, as illustrated in Figure 4.6 and Figure 4.7.

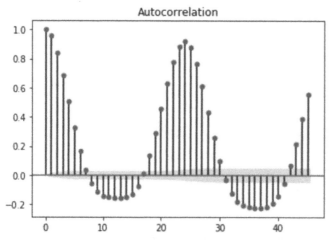

Figure 4.6: Autocorrelation plot results from ts_data_load set with `plot_acf()` function from the statsmodels library

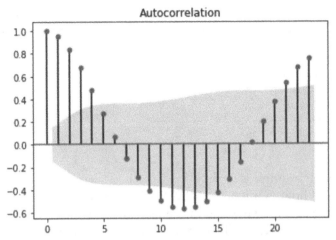

Figure 4.7: Autocorrelation plot results from ts_data_load_subset with `plot_acf()` function from the statsmodels library

As illustrated in Figure 4.6 and Figure 4.7, confidence intervals are drawn as a cone. By default, this is set to a 95 percent confidence interval, suggesting that correlation values outside of this cone are very likely a correlation.

Another important concept to consider is partial correlation function (PACF), which is a conditional correlation. It is the correlation between two variables under the assumption that we consider the values of some other set of variables. In regression, this partial correlation could be found by correlating the residuals from two different regressions.

In a time series data set, the autocorrelation for a value at a time stamp and another value at a prior time stamp consists of both the direct correlation between these two values and indirect correlations. These indirect correlations are a linear function of the correlation of the value under observation, with values at intervening time stamps.

Python supports the PACF function with the `plot_pacf()` from the statsmodels library (`statsmodels.org/stable/generated/statsmodels.graphics.tsaplots.plot_pacf.html`). The example below calculates and plots a partial autocorrelation function for the first 20 lags of our ts_data_load set using the `plot_pacf()` from the statsmodels library:

```
# Import plot_pacf() function
from statsmodels.graphics.tsaplots import plot_pacf

# Plot the pacf function on the ts_data_load dataset
plot_pacf(ts_data_load, lags=20)
pyplot.show()
```

Similarly, the following example calculates and plots a partial autocorrelation function for the first 30 lags of our ts_data_load subset using the `plot_pacf()` from the statsmodels library:

```
# import plot_pacf() function
from statsmodels.graphics.tsaplots import import plot_pacf

# plot the pacf function on the ts_data_load_subset
plot_pacf(ts_data_load_subset, lags=30)
pyplot.show()
```

The x values (which in our example is ts_data_load) can be a series or an array. The argument `lags` shows how many lags of PACF will be plotted. Running these examples creates two 2D plots showing the partial autocorrelation for the first 20 lags and 30 lags respectively, as illustrated in Figure 4.8 and Figure 4.9.

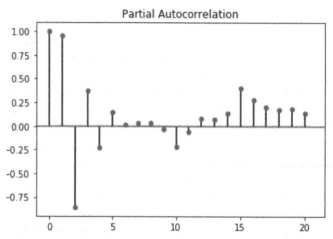

Figure 4.8: Autocorrelation plot results from ts_data set with `plot_pacf()` function from the statsmodels library

The concepts and respective plots of ACF and PACF functions become particularly important when data scientists need to understand and determine the order of autoregressive and moving average time series methods. There are two methods that you can leverage to identify the order of an AR(p) model:

- ▪ The ACF and PACF functions
- ▪ The information criteria

As illustrated in Figure 4.8, ACF is an autocorrelation function that provides you with the information of how much a series is autocorrelated with its lagged values. In simple terms, it describes how well the present value of the series is related with its past values. As we saw in Chapter 3, "Time Series Data Preparation," a time series data set can have components like trend, seasonality, and cyclic patterns. ACF considers all these components while finding correlations.

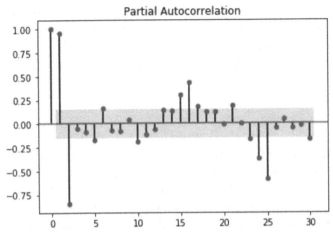

Figure 4.9: Autocorrelation plot results from ts_data_load_subset with `plot_pacf()` function from the statsmodels library

On the other side, PACF is another important function that, instead of finding correlations of present values with lags like ACF, finds correlation of the residuals with the next lag. It is a function that measures the incremental benefit of adding another lag. So if through the PACF function we discover that there is hidden information in the residual that can be modeled by the next lag, we might get a good correlation, and we will keep that next lag as a feature while modeling.

Now let's see why these two functions are important when building an autoregression model. As mentioned at the beginning of this chapter, an autoregression is a model based on the assumption that present values of a time series can be obtained using previous values of the same time series: the present value is a weighted average of its past values.

In order to avoid multicollinear features for time series models, it is necessary to find optimum features or order of the autoregression process using the PACF plot, as it removes variations explained by earlier lags, so we get only the relevant features (Figure 4.8). Notice that we have good positive correlation with the lags up to lag number 6; this is the point where the ACF plot cuts the upper confidence threshold. Although we have good correlation up to the sixth lag, we cannot use all of them as it will create a multicollinearity problem; that's why we turn to the PACF plot to get only the most relevant lags.

In Figure 4.9, we can see that lags up to 6 have good correlation before the plot first cuts the upper confidence interval. This is our p value, which is the order of our autoregression process. We can model the given autoregression process using linear combination of first 6 lags. In Figure 4.9, we can also see that lags up to 1 have good correlation before the plot first cuts the upper confidence interval. This is our p value, the order of our autoregression process. We can then model this autoregression process using the first lag.

The more lag variables you include in your model, the better the model will fit the data; however, this can also represent a risk of overfitting your data. The information criteria adjust the goodness-of-fit of a model by imposing a penalty based on the number of parameters used. There are two popular adjusted goodness-of-fit measures:

- AIC

- BIC

In order to get the information from these two measures, you can use the `summary()` function, the `params` attribute, or the `aic` and `bic` attributes in Python. These information criteria are used to fit several models, each with a different number of parameters, and choose the one with the lowest Bayesian information criterion. For example, if we have an AR(5) model, the lowest information criterion results will denote a value of 5.

Beginning in version 0.11, statsmodels has introduced a new class dedicated to autoregressions (`statsmodels.org/stable/generated/statsmodels.tsa.ar_model.AutoReg.html`), as summarized in Table 4.3.

Table 4.3: Autoregressive class in statsmodels

AUTOREGRESSIVE METHOD	DESCRIPTION AND PROPERTIES
`ar_model.AutoReg(endog, lags[, trend, ...])`	Autoregressive AR-X(p) model
`ar_model.AutoRegResults(model, params, ...)`	Class to hold results from fitting an AutoReg model
`ar_model.ar_select_order(endog, maxlag[, ...])`	Autoregressive AR-X(p) model order selection

The `ar_model.AutoReg` model estimates parameters by applying the following element:

- *A conditional maximum likelihood (CML) estimator,* which is a method that involves the maximization of a conditional log-likelihood function, whereby the parameters treated as known are either fixed by theoretical assumption or, more commonly, replaced by estimates (Lewis-Beck, Bryman, and Liao 2004).

and supporting the following elements:

- *Exogenous regressors* (or independent variables, which are the ones that are having an effect on the values of the target variable) and, more specifically, supporting *ARX models* (autoregressions with exogenous variables). ARX models and related models can also be fitted with the `arima.ARIMA` class and the `SARIMAX` class.

■ *Seasonal effects,* which are defined as systematic and calendar related effects in time series.

Python supports the autoregression modeling using the `AutoReg` model from the statsmodels library (`statsmodels.org/stable/generated/statsmodels.tsa.ar_model.AutoReg.html`), as summarized in Table 4.4.

Table 4.4: Definition and parameters of autoregressive class in statsmodels

CLASS NAME	`statsmodels.tsa.ar_model.autoreg`
Definition	Class that estimates an ARX model using CML estimator.
Parameter	array_like
endog	A 1-d endogenous response variable. The independent variable.
Parameter	{int, list[int]}
Lags	The number of lags to include in the model if an integer or the list of lag indices to include.
Parameter	The trend to include in the model:
trend	'n' - No trend.
: {'n', 'c', 't', 'ct'}	'c' - Constant only.
	't' - Time trend only.
	'ct' - Constant and time trend.
Parameter	bool
seasonal	Flag indicating whether to include seasonal dummies in the model. If seasonal is True and trend includes 'c', then the first period is excluded from the seasonal terms.
Parameter	array_like, optional
exog	Exogenous variables to include in the model. Must have the same number of observations as endog and should be aligned so that endog[i] is regressed on exog[i].
Parameter	{None, int}
hold_back	Initial observations to exclude from the estimation sample. If None, then hold_back is equal to the maximum lag in the model. Set to a non-zero value to produce comparable models with different lag length.

Continues

Table 4.4   *(continued)*

CLASS NAME	statsmodels.tsa.ar_model.autoreg
Parameter	{None, int}
period	The period of the data. Only used if seasonal is True. This parameter can be omitted if using a pandas object for endog that contains a recognized frequency.
Parameter	str
missing	Available options are 'none', 'drop', and 'raise'. If 'none', no nan checking is done. If 'drop', any observations with nans are dropped. If 'raise', an error is raised. Default is 'none'.

We can use this package by first creating the model AutoReg and then calling the fit() function to train it on our data set. Let's see how to apply this package to our ts_data set:

```
# Import necessary libraries
%matplotlib inline
import matplotlib.pyplot as plt
import seaborn as sns
from statsmodels.tsa.ar_model import AutoReg, ar_select_order
from statsmodels.tsa.api import acf, pacf, graphics

# Apply AutoReg model
model = AutoReg(ts_data_load, 1)
results = model.fit()
results.summary()
```

The results.summary() returns a summary of the results for AutoReg model:

AUTOREG MODEL RESULTS			
DEP. VARIABLE:	LOAD	**NO. OBSERVATIONS:**	26304
Model:	AutoReg(1)	**Log Likelihood**	−171640
Method:	Conditional MLE	**S.D. of innovations**	165.1
Date:	Tue, 28 Jan 2020	**AIC**	10.213
Time:	17:05:24	**BIC**	10.214
Sample:	1/1/2012	**HQIC**	10.214
	−2057		

	COEF	STD ERR	Z	P>|Z|	[0.025	0.975]
intercept	144.5181	5.364	26.945	0	134.006	155.03
load.L1	0.9563	0.002	618.131	0	0.953	0.959

ROOTS				
	REAL	IMAGINARY	MODULUS	FREQUENCY
AR.1	1.0457	+0.0000j	1.0457	0

AutoReg supports the same covariance estimators as Ordinary Least-Squares (OLS) models. In the example below, we use `cov_type= "HC0"`, which is White's covariance estimator. The White test is used to detect heteroskedastic errors in regression analysis: the null hypothesis for White's test is that the variances for the errors are equal. While the parameter estimates are the same, all of the quantities that depend on the standard error change (White 1980).

In the following example, I show how to apply covariance estimators `cov_type="HC0"` and output a summary of the results:

```
# Apply covariance estimators cov_type="HC0" and output the summary
res = model.fit(cov_type="HC0")
res.summary()
```

AUTOREG MODEL RESULTS			
DEP. VARIABLE:	LOAD	NO. OBSERVATIONS:	26304
Model:	AutoReg(1)	Log Likelihood	−171640
Method:	Conditional MLE	S.D. of innovations	165.1
Date:	Tue, 28 Jan 2020	AIC	10.213
Time:	20:29:08	BIC	10.214
Sample:	1/1/2012	HQIC	10.214
	−2057		

	COEF	STD ERR	Z	P>|Z|	[0.025	0.975]
intercept	144.5181	5.364	26.945	0	134.006	155.03
load.L1	0.9563	0.002	618.131	0	0.953	0.959

| | | ROOTS | |
REAL	IMAGINARY	MODULUS	FREQUENCY
1.0457	+0.0000j	1.0457	0

By using `plot_predict`, we can visualize the forecasts. Below we produce a large number of forecasts that show the string seasonality captured by the model:

```
# Define figure style, plot package and default figure size
sns.set_style("darkgrid")
pd.plotting.register_matplotlib_converters()

# Default figure size
sns.mpl.rc("figure", figsize=(16, 6))

# Use plot_predict and visualize forecasts
figure = results.plot_predict(720, 840)
```

The preceding code outputs a forecast plot, as illustrated in Figure 4.10.

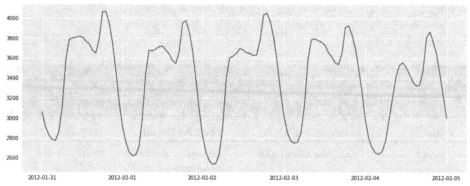

Figure 4.10: Forecast plot generated from ts_data set with `plot_predict()` function from the statsmodels library

`plot_diagnositcs` indicates that the model captures the key features in the data, as shown in the following sample code:

```
# Define default figure size
fig = plt.figure(figsize=(16,9))

# Use plot_predict and visualize forecasts
fig = res.plot_diagnostics(fig=fig, lags=20)
```

The code above outputs four different visualizations, as illustrated in Figure 4.11.

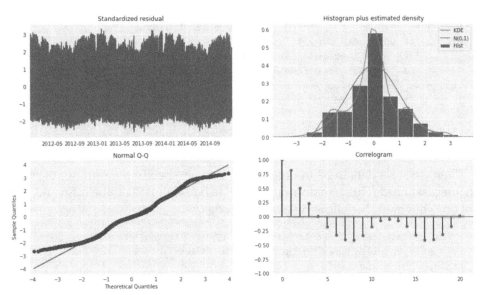

Figure 4.11: Visualizations generated from ts_data set with `plot_diagnositcs()` function from the statsmodels library

Finally, we can test the forecasting capability of the `AutoReg()` function: forecasts are produced using the `predict` method from a results instance. The default produces static forecasts, which are one-step forecasts. Producing multi-step forecasts requires using `dynamic=True`.

In the sample code below, we apply the `predict` method on our ts_data set. Our data preparation for the training set will involve the following steps:

1. Define the start date for the train and test sets.

2. Filter the original data set to include only that time period reserved for the training set.

3. Scale the time series such that the values fall within the interval (0, 1). For this operation, we will use `MinMaxScaler()`; that is an estimator that scales and translates each feature individually such that it is in the given range on the training set—such as between zero and one (scikit-learn.org/ stable/modules/generated/sklearn.preprocessing.MinMaxScaler.html).

In the following sample code, we first define the start date for the train and test data sets:

```
# Define the start date for the train and test sets
train_start_dt = '2014-11-01 00:00:00'
test_start_dt = '2014-12-30 00:00:00'
```

As the second step, we filter the original data set to include only that time period reserved for the training set:

```
# Create train set containing only the model features
train = ts_data_load.copy()[
        (ts_data_load.index >= train_start_dt)
        & (ts_data_load.index < test_start_dt)][['load']]
test = ts_data_load.copy()
        [ts_data_load.index >= test_start_dt][['load']]

print('Training data shape: ', train.shape)
print('Test data shape: ', test.shape)
```

Then we scale data to be in range (0, 1), and we specify the number of steps to forecast ahead. This transformation should be calibrated on the train and test sets, as shown in the following example code:

```
# Scale train data to be in range (0, 1)
from sklearn.preprocessing import MinMaxScaler
scaler = MinMaxScaler()
train['load'] = scaler.fit_transform(train)
train.head()

# Scale test data to be in range (0, 1)
test['load'] = scaler.transform(test)
test.head()
```

We need also to specify the number of steps to forecast ahead, as shown in the example code below:

```
# Specify the number of steps to forecast ahead
HORIZON = 3
print('Forecasting horizon:', HORIZON, 'hours')
```

Now we can create a test data point for each horizon, as illustrated in the following sample code:

```
# Create a test data point for each HORIZON
step.test_shifted = test.copy()

for t in range(1, HORIZON):
        test_shifted['load+'+str(t)] = test_shifted['load']
        .shift(-t, freq='H')

test_shifted = test_shifted.dropna(how='any')
test_shifted.head(5)
```

Finally we can use the `predict` method to make predictions on the test data:

```
%%time
# Make predictions on the test data
training_window = 720

train_ts = train['load']
test_ts = test_shifted

history = [x for x in train_ts]
history = history[(-training_window):]

predictions = list()

for t in range(test_ts.shape[0]):
        model = AutoReg(ts_data_load, 1)
        model_fit = model.fit()
        yhat = model_fit.predict
        predictions.append(yhat)
        obs = list(test_ts.iloc[t])
        # move the training window
        history.append(obs[0])
        history.pop(0)
        print(test_ts.index[t])
        print(t+1, ': predicted =', yhat, 'expected =', obs)
```

In this section of the chapter, you discovered how to build autoregression fore-casts with time series data using Python. In the next section, you will discover the best tools to develop and implement a moving average model with Python.

Moving Average

The moving average technique leverages previous forecast errors in a regres-sion approach to forecast future observations at next time stamps: each future observation can be thought of as a weighted moving average of the previous forecast errors.

Together with the autoregression technique, the moving average approach is an important element of the more general autoregressive moving average and autoregressive integrated moving average models, which present a more sophisticated stochastic configuration.

Moving average models are very similar to autoregressive models: as mentioned in the previous paragraph, autoregressive models represent a linear combination of past observations, while moving average models represent a combination of past error terms. Data scientists usually apply moving average models to their time series data as they are very good at explaining hidden or irregular patterns in the error process directly by fitting a model to the error terms.

Python supports the moving average approach through the `statsmodels.tsa` `.arima_model.ARMA` class (`statsmodels.org/stable/generated/statsmodels` `.tsa.arima_model.ARMA.html`) from statsmodels, by setting the order for the autoregressive model equal to 0 and defining the order of the moving average model in the order argument:

```
MovAvg_Model = ARMA(ts_data_load, order=(0, 1))
```

The statsmodels library offers the definition and parameters for the `ARMA()`class shown in Table 4.5.

Table 4.5: Autoregressive moving average in statsmodels

AUTOREGRESSIVE MOVING AVERAGE	DESCRIPTION AND PROPERTIES
`arima_model.` `ARMA(endog, order[, exog, ...])`	Autoregressive moving average, or ARMA(p,q), model
`arima_model.` `ARMAResults(model, params[, ...])`	Class to hold results from fitting an ARMA model

In this section, you discovered how to build forecasts with moving average models using Python. In the next section, you will discover the best packages and classes to develop and implement autoregressive moving average models with Python.

Autoregressive Moving Average

Autoregressive moving average (ARMA) models have always had a key role in the modeling of time series, as their linear structure contributes to a substantial simplification of linear prediction (Zhang et al. 2015).

An ARMA method consists of two parts:

- An autoregression
- A moving average model

Compared with the autoregressive and moving average models, ARMA models provide the most efficient linear model of stationary time series, since they are capable of modeling the unknown process with the minimum number of parameters (Zhang et al. 2015).

In particular, ARMA models are used to describe weekly stationary stochastic time series in terms of two polynomials. The first of these polynomials is for autoregression, the second for the moving average. Often this method is referred to as the ARMA(p,q) model, in which:

- ▪ p stands for the order of the autoregressive polynomial, and
- ▪ q stands for the order of the moving average polynomial.

Here we will see how to simulate time series from AR(p), MA(q), and ARMA(p,q) processes as well as fit time series models to data based on insights gathered from the ACF and PACF.

Python supports the implementation of the ARMA model with the `ARMA()` function (`statsmodels.org/stable/generated/statsmodels.tsa.arima_model .ARMA.html`) from statsmodels. The statsmodels library offers the definition and parameters for the `ARMA()` class shown in Table 4.6.

Table 4.6: Definition and parameters of autoregressive moving average class in statsmodels

CLASS NAME	STATSMODELS.TSA.ARIMA_MODEL.ARMA
Definition	Class that estimates autoregressive moving average (p,q) model.
Parameter	array_like
endog	The endogenous (independent) variable.
Parameter	iterable
Order	The (p,q) order of the model for the number of AR parameters and MA parameters to use.
Parameter	array_like, `optional`
exog	An optional array of exogenous variables. This should not include a constant or trend. You can specify this in the fit method.
Parameter	array_like, `optional`
dates	An array-like object of datetime objects. If a pandas object is given for endog or exog, it is assumed to have a DateIndex.
Parameter	str, `optional`
freq	The frequency of the time series. A pandas offset or 'B', 'D', 'W', 'M', 'A', or 'Q'. This is optional if dates are given.

Let's now move to the next section and learn how to leverage autoregression and autoregressive moving average to build an autoregressive integrated moving average model with our ts_data_load set.

Autoregressive Integrated Moving Average

Autoregressive integrated moving average (ARIMA) models are considered a development of the simpler autoregressive moving average (ARMA) models and include the notion of *integration*.

Indeed, autoregressive moving average (ARMA) and autoregressive integrated moving average (ARIMA) present many similar characteristics: their elements are identical, in the sense that both of them leverage a general autoregression AR(p) and general moving average model MA(q). As you previously learned, the AR(p) model makes predictions using previous values in the time series, while MA(q) makes predictions using the series mean and previous errors (Petris, Petrone, and Campagnoli 2009).

The main differences between ARMA and ARIMA methods are the notions of integration and differencing. An ARMA model is a stationary model, and it works very well with stationary time series (whose statistical properties, such as mean, autocorrelation, and seasonality, do not depend on the time at which the series has been observed).

It is possible to stationarize a time series through differencing techniques (for example, by subtracting a value observed at time t from a value observed at time t–1). The process of estimating how many nonseasonal differences are needed to make a time series stationarity is called integration (I) or integrated method.

ARIMA models have three main components, denoted as p, d, q; in Python you can assign integer values to each of these components to indicate the specific ARIMA model you need to apply. These parameters are defined as follows:

- *p* stands for the number of lag variables included in the ARIMA model, also called the *lag order*.

- *d* stands for the number of times that the raw values in a time series data set are differenced, also called the *degree of differencing*.

- *q* denotes the magnitude of the moving average window, also called the *order of moving average*.

In the case that one of the parameters above does not need to be used, a value of 0 can be assigned to that specific parameter, which indicates to not use that element of the model.

Let's now take a look at an extension of the ARIMA model in Python, called SARIMAX, which stands for seasonal autoregressive integrated moving average with exogenous factors. Data scientists usually apply SARIMAX when they have to deal with time series data sets that have seasonal cycles. Moreover, SARIMAX models support seasonality and exogenous factors and, as a consequence, they require not only the *p*, *d*, and *q* arguments that ARIMA requires, but also another set of p, d, and q arguments for the seasonality aspect as well as a parameter called s, which is the periodicity of the seasonal cycle in your time series data set.

Python supports the `SARIMAX()` class with the statsmodels library (`statsmodels .org/dev/generated/statsmodels.tsa.statespace.sarimax.SARIMAX.html`), as summarized in Table 4.7.

Table 4.7: Seasonal auto regressive integrated moving average with exogenous factors in statsmodels

SEASONAL AUTO REGRESSIVE INTEGRATED MOVING AVERAGE	DESCRIPTION AND PROPERTIES
`sarimax.` `SARIMAX(endog[, exog, order, ...])`	Seasonal auto regressive integrated moving average with exogenous regressors model
`sarimax.` `SARIMAXResults(model, params, ...` `[, ...])`	Class to hold results from fitting an SARIMAX model

The statsmodels library offers the definition and parameters for the `SARIMAX()` class shown in Table 4.8.

Table 4.8: Definition and parameters of seasonal auto regressive integrated moving average class in statsmodels

CLASS NAME	STATSMODELS.TSA.STATESPACE.SARIMAX.SARIMAX
Definition	Class that estimates a seasonal autoregressive integrated moving average with exogenous regressors model.
Parameter	array_like
endog	The observed time series process yy.
Parameter	array_like, `optional`
exog	Array of exogenous regressors, shaped nobs x k.
Parameter	`iterable or iterable of iterables, optional`
order	The (p,d,q) order of the model for the number of AR parameters, differences, and MA parameters. d must be an integer indicating the integration order of the process, while p and q may either be integers indicating the AR and MA orders (so that all lags up to those orders are included) or else iterables giving specific AR and / or MA lags to include. Default is an AR(1) model: (1,0,0).

Continues

Table 4.8 *(continued)*

CLASS NAME	STATSMODELS.TSA.STATESPACE.SARIMAX.SARIMAX
Parameter	iterable, optional
seasonal_order	The (p,d,q,s) order of the seasonal component of the model for the AR parameters, differences, MA parameters, and periodicity. *d* must be an integer indicating the integration order of the process, while *p* and *q* may either be integers indicating the AR and MA orders (so that all lags up to those orders are included) or else iterables giving specific AR and / or MA lags to include. *s* is an integer giving the periodicity (number of periods in season); often it is 4 for quarterly data or 12 for monthly data. Default is no seasonal effect.
Parameter	str{'n','c','t','ct'}, iterable, optional
trend	Parameter controlling the deterministic trend polynomial A(t)A(t). Can be specified as a string where 'c' indicates a constant (i.e., a degree zero component of the trend polynomial), 't' indicates a linear trend with time, and 'ct' is both. Can also be specified as an iterable defining the polynomial as in *numpy.poly1d*, where *[1,1,0,1]* would denote a+bt+ct3a+bt+ct3. Default is to not include a trend component.
Parameter	bool, optional
measurement_error	Whether or not to assume the endogenous observations *endog* were measured with error. Default is False.
Parameter	bool, optional
time_varying_ regression	Used when an explanatory variable, *exog*, is provided to select whether or not coefficients on the exogenous regressors are allowed to vary over time. Default is False.
Parameter	bool, optional
mle_regression	Whether or not to use the regression coefficients for the exogenous variables as part of maximum likelihood estimation or through the Kalman filter (i.e., recursive least squares). If *time_ varying_regression* is True, this must be set to False. Default is True.
Parameter	bool, optional
simple_differencing	Whether or not to use partially conditional maximum likelihood estimation. If True, differencing is performed prior to estimation, which discards the first sD+dsD+d initial rows but results in a smaller state-space formulation. See the Notes section for important details about interpreting results when this option is used. If False, the full SARIMAX model is put in state-space form so that all datapoints can be used in estimation. Default is False.

CLASS NAME	STATSMODELS.TSA.STATESPACE.SARIMAX.SARIMAX
Parameter	bool, optional
enforce_stationarity	Whether or not to transform the AR parameters to enforce stationarity in the autoregressive component of the model. Default is True.
Parameter	bool, optional
enforce_invertibility	Whether or not to transform the MA parameters to enforce invertibility in the moving average component of the model. Default is True.
Parameter	bool, optional
hamilton_ representation	Whether or not to use the Hamilton representation of an ARMA process (if True) or the Harvey representation (if False). Default is False.
Parameter	bool, optional
concentrate_scale	Whether or not to concentrate the scale (variance of the error term) out of the likelihood. This reduces the number of parameters estimated by maximum likelihood by one, but standard errors will then not be available for the scale parameter.
Parameter	int, optional
trend_offset	The offset at which to start time trend values. Default is 1, so that if *trend='t'* the trend is equal to 1, 2, . . ., nobs. Typically is only set when the model is already created by extending a previous data set.
Parameter	bool, optional
use_exact_diffuse	Whether or not to use exact diffuse initialization for non-stationary states. Default is False (in which case approximate diffuse initialization is used).

Let's now see how to apply a SARIMAX model to our ts_data_load set. In the following code examples, we demonstrate how to do the following:

- Prepare time series data for training a SARIMAX times series forecasting model

- Implement a simple SARIMAX model to forecast the next HORIZON steps ahead (time $t + 1$ through $t + $ HORIZON) in the time series

- Evaluate the model

- In the following sample code, we import the necessary libraries:

```
# Import necessary libraries
from statsmodels.tsa.statespace.sarimax import SARIMAX
from sklearn.preprocessing import MinMaxScaler
import math
from common.utils import mape
```

At this point, we need to separate our data set into train and test sets. We train the model on the train set. After the model has finished training, we evaluate the model on the test set. We must ensure that the test set covers a later period in time than the training set, to ensure that the model does not gain from information from future time periods.

We will allocate the period 1st September to 31st October 2014 to training set (two months) and the period 1st November 2014 to 31st December 2014 to the test set (two months). Since this is daily consumption of energy, there is a strong seasonal pattern, but the consumption is most similar to the consumption in the recent days. Therefore, using a relatively small window of time for training the data should be sufficient:

```
# Create train set containing only the model features
train = ts_data_load.copy()
        [(ts_data_load.index >= train_start_dt)
        & (ts_data_load.index < test_start_dt)][['load']]
test = ts_data_load.copy()
        [ts_data_load.index >= test_start_dt][['load']]

print('Train data shape: ', train.shape)
print('Test data shape: ', test.shape)
```

Our data preparation for the train and test sets involves also scaling the time series such that the values fall within the interval (0, 1):

```
# Scale train data to be in range (0, 1)
scaler = MinMaxScaler()
train['load'] = scaler.fit_transform(train)
train.head()

# Scale test data to be in range (0, 1)
test['load'] = scaler.transform(test)
test.head()
```

It is important to specify the number of steps to forecast ahead and the order and seasonal order for our SARIMAX model:

```
# Specify the number of steps to forecast ahead
HORIZON = 3
print('Forecasting horizon:', HORIZON, 'hours')
```

We then specify the order and seasonal order for the SARIMAX model, as shown in the following example code:

```
# Define the order and seasonal order for the SARIMAX model
order = (4, 1, 0)
seasonal_order = (1, 1, 0, 24)
```

We are finally able to build and fit the model, as shown in the example code below:

```
# Build and fit the SARIMAX model
model = SARIMAX(endog=train, order=order, seasonal_order=seasonal_order)
results = model.fit()

print(results.summary())
```

We now perform the so-called walk forward validation. In practice, time series models are re-trained each time a new data becomes available. This allows the model to make the best forecast at each time step.

Starting at the beginning of the time series, we train the model on the train data set. Then we make a prediction on the next time step. The prediction is then evaluated against the known value. The training set is then expanded to include the known value, and the process is repeated. (Note that we keep the training set window fixed, for more efficient training, so every time we add a new observation to the training set, we remove the observation from the beginning of the set.)

This process provides a more robust estimation of how the model will perform in practice. However, it comes at the computation cost of creating so many models. This is acceptable if the data is small or if the model is simple but could be an issue at scale.

Walk-forward validation is the gold standard of time series model evaluation and is recommended for your own projects:

```
# Create a test data point for each HORIZON step
test_shifted = test.copy()

for t in range(1, HORIZON):
    test_shifted['load+'+str(t)] = test_shifted['load'].shift
(-t, freq='H')

test_shifted = test_shifted.dropna(how='any')
```

We can make predictions on the test data and use a simpler model (by specifying a different order and seasonal order) for demonstration:

```
%%time
# Make predictions on the test data
training_window = 720

train_ts = train['load']
test_ts = test_shifted
```

Continues

(continued)

```
history = [x for x in train_ts]
history = history[(-training_window):]

predictions = list()

# Let's user simpler model
order = (2, 1, 0)
seasonal_order = (1, 1, 0, 24)

for t in range(test_ts.shape[0]):
    model = SARIMAX(endog=history, order=order, seasonal_order=seasonal_
order)
    model_fit = model.fit()
    yhat = model_fit.forecast(steps = HORIZON)
    predictions.append(yhat)
    obs = list(test_ts.iloc[t])
    # move the training window
    history.append(obs[0])
    history.pop(0)
    print(test_ts.index[t])
    print(t+1, ': predicted =', yhat, 'expected =', obs)
```

Let's compare predictions to actual load:

```
# Compare predictions to actual load
eval_df = pd.DataFrame(predictions,
columns=['t+'+str(t) for t in range(1, HORIZON+1)])
eval_df['timestamp'] = test.index[0:len(test.index)-HORIZON+1]
eval_df = pd.melt(eval_df, id_vars='timestamp',
value_name='prediction', var_name='h')
eval_df['actual'] = np.array(np.transpose(test_ts)).ravel()
eval_df[['prediction', 'actual']] = scaler.inverse_transform(eval_
df[['prediction', 'actual']])
```

It is also helpful to compute the mean absolute percentage error (MAPE), which measures the size of the error in percentage terms over all predictions, as shown in the following example:

```
# Compute the mean absolute percentage error (MAPE)
if(HORIZON > 1):
    eval_df['APE'] = (eval_df['prediction'] -
        eval_df['actual']).abs() / eval_df['actual']
    print(eval_df.groupby('h')['APE'].mean())
```

In order to look at the MAPE results, we print both the one-step forecast MAPE result and the multi-step forecast MAPE result:

```
# Print one-step forecast MAPE
print('One step forecast MAPE: ', (mape(eval_df[eval_df['h']
```

```
== 't+1']['prediction'],
eval_df[eval_df['h'] == 't+1']['actual']))*100, '%')

# Print multi-step forecast MAPE
print('Multi-step forecast MAPE: ',
mape(eval_df['prediction'], eval_df['actual'])*100, '%')
```

In this section, you learned how to build, train, test, and validate a SARIMAX model for your forecasting solution. Let's now move to the next section and learn how to leverage Automated ML for time series forecasting.

Automated Machine Learning

In this section, you will learn how to train a time series forecasting regression model using Automated ML in Azure Machine Learning (aka.ms/AzureMLservice). Designing a forecasting model is similar to setting up a typical regression model using Automated ML (aka.ms/AutomatedML); however, it is important to understand what configuration options and pre-processing steps exist for time series data: the most important difference between a forecasting regression task and regression task within Automated ML is including a variable in your data set as a designated time stamp column (aka.ms/AutomatedML).

The following examples in Python show you how to do the following:

■ Prepare your data for time series forecasting with Automated ML

■ Configure specific time series parameters in an AutoMLConfig object using 'AutoMLConfig'

■ Train the model using AmlCompute, which is a managed-compute infrastructure that allows you to easily create a single or multi-node compute

■ Explore the engineered features and results

If you are using a cloud-based Azure Machine Learning compute instance, you are ready to start coding by using either the Jupyter notebook or JupyterLab experience. You can find more information on how to configure a development environment for Azure Machine Learning at docs.microsoft.com/en-us/azure/machine-learning/how-to-configure-environment.

Otherwise, you need to go through the configuration steps to establish your connection to the Azure Machine Learning workspace. You can visit the following links to configure your Azure Machine Learning workspace and learn how to use Jupyter notebooks on Azure Machine Learning:

■ Aka.ms/AzureMLConfiguration

■ Aka.ms/AzureMLJupyterNotebooks

In the following example, we set up important resources and packages to run Automated ML on Azure Machine Learning and to manage eventual waning messages in the notebook:

```
# Import resources and packages for Automated ML and time series
forecasting
import logging
from sklearn.metrics import mean_absolute_error, mean_squared_error,
r2_score
from matplotlib import pyplot as plt
import pandas as pd
import numpy as np
import warnings
import os
import azureml.core
from azureml.core import Experiment, Workspace, Dataset
from azureml.train.automl import AutoMLConfig
from datetime import datetime


# manage warning messages
warnings.showwarning = lambda *args, **kwargs: None
```

As part of the setup you have already created an Azure ML workspace object (Aka.ms/AzureMLConfiguration). Moreover, for Automated ML, you will also need to create an Experiment object, which is a named object in a workspace used to run machine learning experiments:

```
# Select a name for the run history container in the workspace
experiment_name = 'automatedML-timeseriesforecasting'

experiment = Experiment(ws, experiment_name)
output = {}
output['SDK version'] = azureml.core.VERSION
output['Subscription ID'] = ws.subscription_id
output['Workspace'] = ws.name
output['Resource Group'] = ws.resource_group
output['Location'] = ws.location
output['Run History Name'] = experiment_name
pd.set_option('display.max_colwidth', -1)
outputDf = pd.DataFrame(data = output, index = [''])
outputDf.T
```

A compute target is required to execute a remote run of your Automated ML experiment. Azure Machine Learning compute is a managed-compute infrastructure that allows you to create a single- or multi-node compute.

In this tutorial, we create AmlCompute as your training compute resource. To learn more about how to set up and use compute targets for model training, you can visit `docs.microsoft.com/en-us/azure/machine-learning/how-to-set-up-training-targets`. It is important to keep in mind that, as with other Azure services, there are limits on certain resources (such as AmlCompute) associated with the Azure Machine Learning service. Please read this article on the default limits and how to request more quota: `aka.ms/AzureMLQuotas`.

Let's now import AmlCompute and the compute target, as well as select a name for our cluster and check whether the compute target already exists in the workspace:

```
# Import AmlCompute and ComputeTarget for the experiment
from azureml.core.compute import ComputeTarget, AmlCompute
from azureml.core.compute_target import ComputeTargetException

# Select a name for your cluster
amlcompute_cluster_name = "tsf-cluster"

# Check if that cluster does not exist already in the workspace
try:
    compute_target = ComputeTarget(workspace=ws, name=amlcompute_
cluster_name)
    print('Found existing cluster, use it.')
except ComputeTargetException:
    compute_config = AmlCompute.provisioning_configuration
(vm_size='STANDARD_DS12_V2', max_nodes=6)
    compute_target = ComputeTarget.create
(ws, amlcompute_cluster_name, compute_config)
compute_target.wait_for_completion(show_output=True)
```

Let's now define and prepare our time series data for forecasting with Automated ML. For this Automated ML example, we use a New York City energy demand dataset (`mis.nyiso.com/public/P-58Blist.htm`): the data set includes consumption data from New York City stored in a tabular format and includes energy demand and numerical weather features at an hourly frequency. The purpose of this experiment is to predict the energy demand in New York City for the next 24 hours by building a forecasting solution that leverages historical energy data from the same geographical region.

In case you are interested in exploring additional public data sets and features (such as weather, satellite imagery, socioeconomic data) and add them to this energy data set to improve the accuracy of your machine learning models, I recommend checking out the Azure Open Datasets catalog: `aka.ms/AzureOpenDatasetsCatalog`.

For our Automated ML experiment, we need to identify the target column, which represents the target variable that we want to forecast. The time column is our time stamp column that defines the temporal structure of our data set.

Finally, there are two additional columns, *temp* and *precip*, that represent two additional numerical weather variables that we can include in our forecasting experiment:

```
# Identify the target and time column names in our data set
target_column_name = 'demand'
time_column_name = 'timeStamp'
```

At this point we can load the data set using the TabularDataset class (docs.microsoft.com/en-us/python/api/azureml-core/azureml.data.tabulardataset?view=azure-ml-py). To get started working with a tabular dataset, see aka.ms/tabulardataset-samplenotebook.

```
# load the data set using the TabularDataset class
ts_data = Dataset.Tabular.from_delimited_files
(path = "https://automlsamplenotebookdata.blob.core.windows.net
/automl-sample-notebook-data/nyc_energy.csv")
.with_timestamp_columns(fine_grain_timestamp=time_column_name)
ts_data.take(5).to_pandas_dataframe().reset_index(drop=True)
```

This energy data set is missing energy demand values for all datetimes later than August 10, 2017, 5:00. Below, we reduce and delete the rows containing these missing values from the end of the data set:

```
# Delete a few rows of the data set due to large number of NaN values
ts_data = ts_data.time_before(datetime(2017, 10, 10, 5))
```

The first split we make is into train and test sets: data before and including August 8, 2017, 5:00 will be used for training, and data after will be used for testing:

```
# Split the data set into train and test sets
train = ts_data.time_before(datetime(2017, 8, 8, 5), include_
boundary=True)
train.to_pandas_dataframe().reset_index(drop=True)
.sort_values(time_column_name).tail(5)

test = ts_data.time_between(datetime(2017, 8, 8, 6), datetime(2017, 8,
10, 5))
test.to_pandas_dataframe().reset_index(drop=True).head(5)
```

The forecast horizon is the number of future time stamps that the model should predict. In this example, we set the horizon to 24 hours:

```
# Define the horizon length to 24 hours
max_horizon = 24
```

For forecasting tasks, Automated ML uses pre-processing and estimation steps that are specific to time series data (`aka.ms/AutomatedML`). It first detects time series sample frequency (for example, hourly, daily, weekly) and creates new records for absent time points to make the series continuous. Then it imputes missing values in the target (via forward-fill) and feature columns (using median column values) and creates grain-based features to enable fixed effects across different series. Finally, it creates time-based features to assist in learning seasonal patterns and encodes categorical variables to numeric quantities. To learn more about this process visit the page `aka.ms/AutomatedML`.

The AutoMLConfig object defines the settings and data necessary for an Automated ML task: data scientists need to define standard training parameters like task type, number of iterations, training data, and number of cross-validations. For forecasting tasks, there are additional parameters that must be set that affect the experiment. Table 4.9 summarizes each parameter and its usage.

Table 4.9: Automated ML parameters to be configured with the AutoML Config class

PROPERTY	DESCRIPTION
`Task`	forecasting
`primary_metric`	This is the metric that you want to optimize.
	Forecasting supports the following primary metrics: ▪ `spearman_correlation` ▪ `normalized_root_mean_squared_error` ▪ `r2_score` ▪ `normalized_mean_absolute_error`
`blocked_models`	Models in `blocked_models` won't be used by AutoML.
`experiment_timeout_hours`	Maximum amount of time in hours that the experiment can take before it terminates.
`training_data`	The training data to be used within the experiment.
`label_column_name`	The name of the label column.
`compute_target`	The remote compute for training.
`n_cross_validations`	Number of cross-validation splits. Rolling Origin Validation is used to split time series in a temporally consistent way.
`enable_early_stopping`	Flag to enable early termination if the score is not improving in the short term.
`time_column_name`	The name of your time column.
`max_horizon`	The number of periods out you would like to predict past your training data. Periods are inferred from your data.

The code below shows how to set those parameters in Python. Specifically, you will use the `blacklist_models` parameter to exclude some models. You can choose to remove models from the `blacklist_models` list and increase the `experiment_timeout_hours` parameter value to see your Automated ML results:

```python
# Automated ML configuration
automl_settings = {
    'time_column_name': time_column_name,
    'max_horizon': max_horizon,
}

automl_config = AutoMLConfig(task='forecasting',
                    primary_metric='normalized_root_mean_
                    squared_error',
                    blocked_models =
                      ['ExtremeRandomTrees',
                       'AutoArima', 'Prophet'],
                    experiment_timeout_hours=0.3,
                    training_data=train,
                    label_column_name=target_column_name,
                    compute_target=compute_target,
                    enable_early_stopping=True,
                    n_cross_validations=3,
                    verbosity=logging.INFO,
                 **automl_settings)
```

We now call the submit method on the experiment object and pass the run configuration. Depending on the data and the number of iterations, this can run for a while. One may specify `show_output = True` to print currently running iterations to the console:

```python
# Initiate the remote run
remote_run = experiment.submit(automl_config, show_output=False)
remote_run
```

Below we select the best model from all the training iterations using `get_output` method:

```python
# Retrieve the best model
best_run, fitted_model = remote_run.get_output()
fitted_model.steps
```

For time series task types in Automated ML, you can view details from the feature engineering process. The following code shows each raw feature along with the following attributes:

▪ Raw feature name

▪ Number of engineered features formed out of this raw feature

▪ Type detected

- Whether feature was dropped
- List of feature transformations for the raw feature

```
# Get the featurization summary as a list of JSON
featurization_summary = fitted_model.named_steps['timeseriestransformer']
.get_featurization_summary()

# View the featurization summary as a pandas dataframe
pd.DataFrame.from_records(featurization_summary)
```

Now that we have retrieved the best model for our forecasting scenario, it can be used to make predictions on test data. First, we need to remove the target values from the test set:

```
# Make predictions using test data
X_test = test.to_pandas_dataframe().reset_index(drop=True)
y_test = X_test.pop(target_column_name).values
```

For forecasting, we will use the `forecast` Python function, as illustrated in the following sample code:

```
# Apply the forecast function
y_predictions, X_trans = fitted_model.forecast(X_test)
```

To evaluate the accuracy of the forecast, we compare against the actual energy values and we use the MAPE metric for evaluation. We also use the `align_outputs` function in order to line up the output explicitly to the input:

```
# Apply the align_outputs function
# in order to line up the output explicitly to the input
from forecasting_helper import align_outputs
from azureml.automl.core._vendor.automl.client.core.common import
metrics
from automl.client.core.common import constants

ts_results_all = align_outputs
(y_predictions, X_trans, X_test, y_test, target_column_name)

# Compute metrics for evaluation
scores = metrics.compute_metrics_regression(
    ts_results_all['predicted'],
    ts_results_all[target_column_name],
    list(constants.Metric.SCALAR_REGRESSION_SET),
    None,
    None,
    None)
```

In this section of Chapter 4, you learned how to leverage Automated ML for your time series scenario. To learn more how to use Automated ML for classification, regression, and forecasting scenarios, you can visit `aka.ms/AutomatedML`.

Conclusion

In this chapter, you learned how to train a time series forecasting regression model using autoregressive methods and Automated ML in Azure Machine Learning.

Specifically, we looked at the following methods:

- *Autoregression* – This time series technique assumes that future observations at next time stamp are related to the observations at prior time stamps through a linear relationship.

- *Moving Average* – This time series technique leverages previous forecast errors in a regression approach to forecast future observations at next time stamps. Together with the autoregression technique, the moving average approach is a crucial element of the more general autoregressive moving average and autoregressive integrated moving average models, which present a more sophisticated stochastic configuration.

- *Autoregressive Moving Average* – This time series technique assumes that future observations at next time stamp can be represented as a linear function of the observations and residual errors at prior time stamps.

- *Autoregressive Integrated Moving Average* – This time series technique assumes that future observations at next time stamp can be represented as a linear function of the differenced observations and residual errors at prior time stamps.

- *Automated ML* – This time series technique iterates through a portfolio of different machine learning algorithms for time series forecasting while performing best model selection, hyperparameters tuning, and feature engineering for your scenario.

In the next chapter we will learn how to use and apply deep learning methods to time series scenarios.

## CHAPTER 5

Introduction to Neural Networks for Time Series Forecasting

As we discussed in the previous chapters, the ability to accurately forecast a sequence into the future is critical in many industries: finance, supply chain, and manufacturing are just a few examples. Classical time series techniques have served this task for decades, but now deep learning methods—similar to those used in computer vision and automatic translation—have the potential to revolutionize time series forecasting as well.

Due to their applicability to many real-life problems, such as fraud detection, spam email filtering, finance, and medical diagnosis, and their ability to produce actionable results, deep learning neural networks have gained a lot of attention in recent years. Generally, deep learning methods have been developed and applied to univariate time series forecasting scenarios, where the time series consists of single observations recorded sequentially over equal time increments (Lazzeri 2020).

For this reason, they have often performed worse than naïve and classical forecasting methods, such as autoregressive integrated moving average (ARIMA). This has led to a general misconception that deep learning models are inefficient in time series forecasting scenarios, and many data scientists wonder whether it's really necessary to add another class of methods, like convolutional neural networks or recurrent neural networks, to their time-series toolkit (Lazzeri 2020).

In this chapter, we discuss some of the practical reasons data scientists may still think about deep learning when they build time series forecasting solutions. Specifically, we will take a closer look at the following important topics:

▪ *Reasons to Add Deep Learning to Your Time Series Toolkit* – Deep learning neural networks are able to automatically learn arbitrary complex mappings from inputs to outputs and support multiple inputs and outputs. These are powerful features that offer a lot of promise for time series forecasting, particularly on problems with complex-nonlinear dependencies, multivalent inputs, and multi-step forecasting. These features, along with the capabilities of more modern neural networks, may offer great promise, such as the automatic feature learning provided by convolutional neural networks and the native support for sequence data in recurrent neural networks. In this section, you will discover the promised capabilities of deep learning neural networks for time series forecasting. Specifically, we will discuss the following deep learning methods' capabilities:

  ▪ Deep learning neural networks are capable of automatically learning and extracting features from raw and imperfect data.

  ▪ Deep learning supports multiple inputs and outputs

  ▪ Recurrent neural networks—specifically long short-term memory (LSTM) networks—are good at extracting patterns in input data that span over relatively long sequences.

▪ *Recurrent Neural Networks for Time Series Forecasting* – In this section I will introduce a very popular type of artificial neural networks: recurrent neural networks, also known as RNNs. Recurrent neural networks are a class of neural networks that allow previous outputs to be used as inputs while having hidden states.

▪ *How to Develop GRUs and LSTMs for Time Series Forecasting* – In this final section of Chapter 5, you will learn how to prepare your data for deep learning models and develop GRU models for a range of time series forecasting problems.

Reasons to Add Deep Learning to Your Time Series Toolkit

The goal of machine learning is to find features to train a model that transforms input data (such as pictures, time series, or audio) to a given output (such as captions, price values, or transcriptions). Deep learning is a subset of machine learning algorithms that learn to extract these features by representing input data as vectors and transforming them with a series of linear algebra operations into a given output. In order to clarify further the difference between deep learning and machine learning, let's start by defining each of these two fields of study separately:

- *Machine learning* is the practice of using an algorithm to break up data, learn from it, and then use this data to make some predictions about a certain phenomenon. The learning process is generally based on the following steps:

  1. We feed our algorithm with data.

  2. We then use this data to teach our model how to learn from previous observations.

  3. We run a test to check if our model has learned enough from previous observations and we evaluate its performance.

  4. If the model is performing well (based on our expectations and requirements), we deploy and push it into production, making it available to other stakeholders in the organization or outside the business.

  5. Finally, we consume our deployed model to perform a certain automated predictive task.

- *Deep learning* is a subset of machine learning. Deep learning algorithms are specific types of machine learning algorithms that are based on artificial neural networks. In this case, the learning process is based on the same steps as those for machine learning, but it is called deep because the structure of algorithm is based on artificial neural networks that consist of multiple input, output, and hidden layers, containing units that transform the input data into an information that the next layer can use to perform a certain automated predictive task once the deep learning model is deployed.

It is important to compare the two techniques and understand the main differences. The key differences between machine learning and deep learning are summarized in Table 5.1.

Table 5.1: Key differences between machine learning and deep learning

CHARACTERISTIC CATEGORY	ALL MACHINE LEARNING	ONLY DEEP LEARNING
Number of data points	Can use small amounts of data to make predictions.	Needs to use large amounts of training data to make predictions.
Hardware dependencies	Can work on low-end machines. It doesn't need a large amount of computational power.	Depends on high-end machines. It inherently does a large number of matrix multiplication operations. A GPU can efficiently optimize these operations.

Continues

Table 5.1 *(continued)*

CHARACTERISTIC CATEGORY	ALL MACHINE LEARNING	ONLY DEEP LEARNING
Featurization process	Requires features to be accurately identified and created by users.	Learns high-level features from data and creates new features by itself.
Learning approach	Divides the learning process into smaller steps. It then combines the results from each step into one output.	Moves through the learning process by resolving the problem on an end-to-end basis.
Execution time	Takes comparatively little time to train, ranging from a few seconds to a few hours.	Usually takes a long time to train because a deep learning algorithm involves many layers.
Output	The output is usually a numerical value, like a score or a classification.	The output can have multiple formats, like a text, a score, or a sound.

Source: `aka.ms/deeplearningvsmachinelearning`

Data scientists then evaluate whether the output is what they expected, using an equation called loss function. The goal of the process is to use the result of the loss function from each training input to guide the model to extract features that will result in a lower loss value on the next pass (Lazzeri 2019a). Deep learning neural networks have three main intrinsic capabilities:

- Deep learning neural networks are capable of automatically learning and extracting features from raw and imperfect data.

- Deep learning supports multiple inputs and outputs.

- Recurrent neural networks—specifically long short-term memory (LSTM) networks and gated recurrent units (GRUs)—are good at extracting patterns in input data that span over relatively long sequences.

Thanks to these three characteristics, they can offer a lot of help when data scientists deal with more complex but still very common problems, such as time series forecasting (Lazzeri 2019a).

Deep Learning Neural Networks Are Capable of Automatically Learning and Extracting Features from Raw and Imperfect Data

Time series is a type of data that measures how things change over time. In time series, time isn't just a metric, but a primary axis. This additional dimension represents both an opportunity and a constraint for time series data because it provides a source of additional information but makes time series problems more challenging as specialized handling of the data is required (Lazzeri 2019a).

Moreover, this temporal structure can carry additional information, like trends and seasonality, that data scientists need to deal with in order to make their time series easier to model with any type of classical forecasting methods. Neural networks can be useful for time series forecasting problems by eliminating the immediate need for massive feature engineering processes, data scaling procedures, and making the data stationary by differencing.

In real-work time series scenarios—for example, weather forecasting, air quality and traffic flow forecasting, and forecasting scenarios based on streaming IoT devices like geo-sensors—irregular temporal structures, missing values, heavy noise, and complex interrelationships between multiple variables present limitations for classical forecasting methods.

These techniques typically rely on clean, complete data sets in order to perform well; missing values, outliers, and other imperfect features are generally unsupported. Speaking of more artificial and perfect data sets, classical forecasting methods are based on the assumption that a linear relationship and a fixed temporal dependence exist among variables of a data set, and this assumption by default excludes the possibility of exploring more complex, and probably more interesting, relationships among variables.

Data scientists must make subjective judgments when preparing the data for classical analysis—like the lag period used to remove trends—which is time consuming and introduces human biases to the process. On the contrary, neural networks are robust to noise in input data and in the mapping function and can even support learning and prediction in the presence of missing values. Convolutional neural networks (CNNs) are a category of neural networks that have proven very effective in areas such as image recognition and classification. CNNs have been successful in identifying faces, objects, and traffic signs, apart from powering vision in robots and self-driving cars. CNNs derive their name from the "convolution" operator (Lazzeri 2019a).

The primary purpose of convolution in the case of CNNs is to extract features from the input image. Convolution preserves the spatial relationship between pixels by learning image features, using small squares of input data. In other words, the model learns how to automatically extract the features from the raw data that are directly useful for the problem being addressed. This is called representation learning, and the CNN achieves this in such a way that the features are extracted regardless of how they occur in the data, so-called transform or distortion invariance. The ability of CNNs to learn and automatically extract features from raw input data can be applied to time series forecasting problems.

A sequence of observations can be treated like a one-dimensional image that a CNN model can read and refine into the most relevant elements. This capability of CNNs has been demonstrated to great effect on time series classification tasks such as indoor movement prediction, using wireless sensor strength data to predict the location and motion of subjects within a building (Lazzeri 2019a).

Deep Learning Supports Multiple Inputs and Outputs

Real-world time series forecasting is challenging for several reasons, such as having multiple input variables, the requirement to predict multiple time steps, and the need to perform the same type of prediction for multiple physical sites. Deep learning algorithms can be applied to time series forecasting problems and offer benefits such as the ability to handle multiple input variables with noisy complex dependencies (Lazzeri 2019a).

Specifically, neural networks can be configured to support an arbitrary but fixed number of inputs and outputs in the mapping function. This means that neural networks can directly support multivariate inputs, providing direct support for multivariate forecasting. A univariate time series, as the name suggests, is a series with a single time-dependent variable. For example, we want to predict next energy consumption in a specific location: in a univariate time series scenario, our dataset will be based on two variables, time values and historical energy consumption observations (Lazzeri 2019a).

A multivariate time series has more than one time-dependent variable. Each variable not only depends on its past values but also has some dependency on other variables. This dependency is used for forecasting future values. Let's consider the above example again.

Now suppose our data set includes weather data, such as temperature values, dew point, wind speed, and cloud cover percentage along with the energy consumption value for the past four years. In this case, there are multiple variables to be considered to optimally predict energy consumption value. A series like this would fall under the category of multivariate time series.

With neural networks, an arbitrary number of output values can be specified, offering direct support for more complex time series scenarios that require multivariate forecasting and even multi-step forecast methods. There are two main approaches that deep learning methods can be used to make multi-step forecasts:

- *Direct*, where a separate model is developed to forecast each forecast lead time
- *Recursive*, where a single model is developed to make one-step forecasts, and the model is used recursively where prior forecasts are used as input to forecast the subsequent lead time

The recursive approach can make sense when forecasting a short contiguous block of lead times, whereas the direct approach may make more sense when forecasting discontiguous lead times. The direct approach may be more appropriate when we need to forecast a mixture of multiple contiguous and discontiguous lead times over a period of a few days; such is the case, for example, with air pollution forecasting problems or for anticipatory shipping forecasting, used to predict what customers want and then ship the products automatically (Lazzeri 2019a).

Key to the use of deep learning algorithms for time series forecasting is the choice of multiple input data. We can think about three main sources of data that can be used as input and mapped to each forecast lead time for a target variable:

- *Univariate data*, such as lag observations from the target variable that is being forecasted.

- *Multivariate data*, such as lag observations from other variables (for example, weather and targets in case of air pollution forecasting problems).

- *Metadata*, such as data about the date or time being forecast. This type of data can provide additional insight into historical patterns, helping create richer data sets and more accurate forecasts (Brownlee 2017).

Recurrent Neural Networks Are Good at Extracting Patterns from Input Data

Recurrent neural networks (RNNs) were created in the 1980s but have just recently been gaining popularity due to increased computational power from graphic processing units. They are especially useful with sequential data because each neuron or unit can use its internal memory to maintain information about the previous input. An RNN has loops in it that allow information to be carried across neurons while reading in input.

However, a simple recurrent network suffers from a fundamental problem of not being able to capture long-term dependencies in a sequence. This is a major reason RNNs faded out from practice until some great results were achieved with using a LSTM unit inside the neural network. Adding the LSTM to the network is like adding a memory unit that can remember context from the very beginning of the input (Lazzeri 2019a).

LSTM neural networks are a particular type of RNN that have some internal contextual state cells that act as long-term or short-term memory cells. The output of the LSTM network is modulated by the state of these cells. This is a very important property when we need the prediction of the neural network to depend on the historical context of inputs rather than only on the very last input. They are a type of neural network that adds native support for input data comprising sequences of observations. The addition of sequence is a new dimension to the function being approximated. Instead of mapping inputs to outputs alone, the network can learn a mapping function for the inputs over time to an output.

The example of video processing can be very effective when we need to understand how LSTM networks work: in a movie, what happens in the current frame is heavily dependent upon what was in the last frame. Over a period of time, an LSTM network tries to learn what to keep and how much to keep from the past and how much information to keep from the present state, which makes it powerful compared to other types of neural networks (Lazzeri 2019a).

This capability can be used in any time series forecasting context, where it can be extremely helpful to automatically learn the temporal dependence from the data. In the simplest case, the network is shown one observation at a time from a sequence and can learn which prior observations are important and how they are relevant to forecasting. The model both learns a mapping from inputs to outputs and learns what context from the input sequence is useful for the mapping and can dynamically change this context as needed.

Not surprisingly, this approach has often been used in the finance industry to build models that forecast exchange rates, based on the idea that past behavior and price patterns may affect currency movements and can be used to predict future price behavior and patterns. On the other hand, there are downsides that data scientists need to be careful about while implementing neural network architectures. Large volumes of data are required, and models require hyperparameter tuning and multiple optimization cycles.

In this section, we discussed the three neural networks capabilities that can offer a lot of help when data scientists deal with more complex but still very common problems, like time series forecasting. In the next section, we will deep dive on some fundamental concepts to understand better how neural networks work for time series forecasting (Lazzeri 2019a).

Recurrent Neural Networks for Time Series Forecasting

Recurrent neural network (RNN), also known as autoassociative or feedback network, belongs to a class of artificial neural networks where connections between units form a directed cycle. This creates an internal state of the network, which allows it to exhibit dynamic temporal behavior (Poznyak, Oria, and Poznyak 2018).

RNN can leverage their internal memory to handle sequences of inputs and, instead of mapping inputs to outputs alone, it is capable of leveraging a mapping function for the inputs over *time* to an output. RNNs have shown to achieve the state-of-the-art results in many applications with time series or sequential data, including machine translation and speech recognition.

In particular, LSTM is a type of RNN architecture that performs particularly well on different temporal processing tasks, and LSTM networks are able to address the issue of large time lags in input data successfully. LSTM networks have several nice properties such as strong prediction performance as well as the ability to capture long-term temporal dependencies and variable-length observations (Che et al. 2018).

Exploiting the power of customized RNN models along with LSTM models is a promising venue to effectively model time series data: in the next few sections of this chapter, you will see how RNNs, LSTMs, and Python can help data scientists build accurate models for their time series forecasting solutions.

Recurrent Neural Networks

RNNs are neural networks with hidden states and loops, allowing information to persist over time. In this section, I'll first introduce the concept of *recurrent units* and how their memory works, and finally, I'll explain how they can be used to handle sequence data such as time series.

Different types of neural networks (such as feed forward networks) are based on the idea of learning during training from context and history to produce predictions: RNNs use the idea of hidden state (or memory) in order to be able to generate an outcome or prediction by updating each neuron into a new computational unit that is able to remember what it has seen before (Bianchi et al. 2018).

This memory is preserved inside the unit, in an array or in a vector, so that when the unit reads an input, it also processes the content of the memory, combining the information. By using both (the knowledge from the input and the knowledge from the memory), the unit is now capable of making a prediction (Y) and updating the memory itself, as illustrated in Figure 5.1.

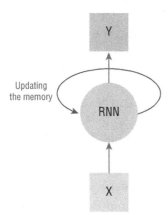

Figure 5.1: Representation of a recurrent neural network unit

RNN units are described as recurrent units because the type of dependence of the current value on the previous event is *recurrent* and can be thought of as multiple copies of the same node, each transmitting a recurrent message to a successor. Let's visualize this recursive relationship in Figure 5.2.

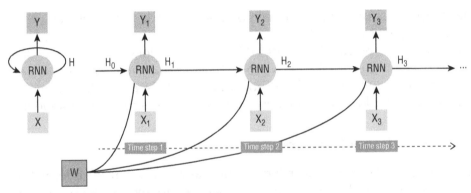

Figure 5.2: Recurrent neural network architecture

In Figure 5.2, an RNN architecture is represented. RNN has an internal hidden state, denoted as H, which can be fed back to the network. In this figure, the RNN processes input values X and produces output values Y. The hidden stat, denoted as H, allows the information to be forwarded from one node of the network to the next node, organized as sequences or a vector. That vector now has information on the current input and previous inputs. The vector goes through the *tanh activation*, and the output is the new hidden state. The tanh activation is used to regulate the values flowing through the network and always keeps the values between –1 and 1. Because of this structure, RNNs can represent a good modeling choice to try to solve a variety of sequential problems, such as time series forecasting, speech recognition, or image captioning (Bianchi 2018).

As you can notice, in Figure 5.2, there is an additional element denoted as W. W indicates that each unit has three sets of weights, one for the inputs (X), another for the outputs of the previous time step (H), and the other for the output of the current time step (Y). Those weight values are determined by the training process and can be identified by applying a popular optimization technique called *gradient descent*.

In simple words, a gradient is the slope of a function that can be defined as the partial derivatives of a set of parameters with respect to its inputs. Gradients are values used to update a neural network's weights: the function represents the problem that we want to solve using neural networks and selecting an appropriate choice of parameters to update those neural networks weights. In order to calculate a gradient descent, we then need to calculate the loss function and its derivative with respect to the weights. We start with a random point on the function and move in the negative direction of the gradient of the function to reach a point where the value of the function is minimum (Zhang et al. 2019).

In Figure 5.2, we seem to be applying the same weights to different items in the input series. This means that we are sharing parameters across inputs. If we are not able share parameters across inputs, then an RNN becomes like an ordinary neural network where each input node requires weights of their own.

RNNs instead can leverage their hidden state property that ties one input to the next one and combines this input connection in a serial input.

Despite their great potential, RNNs have a few limitations, as they suffer from short-term memory. For example, if an input sequence is long enough, they are not able to transfer information from earlier time steps to later ones, and they often leave out important information from the beginning of the sequence during the backpropagation process. We refer to this problem as the *vanishing gradient problem*, as illustrated in Figure 5.3.

During backpropagation, recurrent neural networks suffer from this problem because the gradient decreases as it backpropagates through time. If a gradient value becomes extremely small, it is not able to contribute to the network learning process. LSTM solves this problem by introducing changes to RNN architecture on how they compute outputs and hidden state using the inputs: specifically, LSTMs introduce additional elements in the architecture, called *gates* and *cell state*. There are variants of LSTMs, and we will discuss them in more detail in the next section.

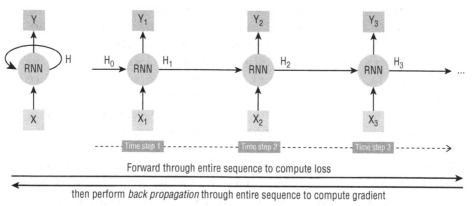

Figure 5.3: Back propagation process in recurrent neural networks to compute gradient values

Long Short-Term Memory

LSTMs are capable of learning long-term dependencies, which is a useful capability when you need to model time series data. As mentioned in the previous section, LSTMs help preserve the error that can be backpropagated through time and layers, without risk of losing important information: LSTMs have internal mechanisms called gates and cell state that can regulate the flow of information (Zhang et al. 2019).

The cell makes decisions about what, how much, and when to store and release information: they learn when to allow information to enter, leave, or be deleted through the iterative process of making guesses, backpropagating error, and adjusting weights via gradient descent. Figure 5.4 illustrates how information flows through a cell and is controlled by different gates.

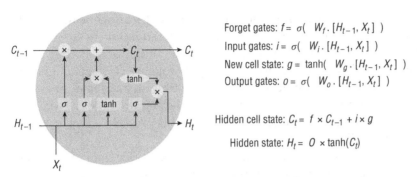

Forget gates: $f = \sigma(\ W_f . [H_{t-1}, X_t]\ )$

Input gates: $i = \sigma(\ W_i . [H_{t-1}, X_t]\ )$

New cell state: $g = \tanh(\ W_g . [H_{t-1}, X_t]\ )$

Output gates: $o = \sigma(\ W_o . [H_{t-1}, X_t]\ )$

Hidden cell state: $C_t = f \times C_{t-1} + i \times g$

Hidden state: $H_t = O \times \tanh(C_t)$

Figure 5.4: Backpropagation process in recurrent neural networks to compute gradient values

In Figure 5.4, it is important to note that LSTM memory cells leverage both *addition* and *multiplication* in the transformation of input and transfer of information. The addition step is essentially the secret of LSTMs: it preserves a constant error when it must be backpropagated at depth. Instead of influencing the next cell state by multiplying its current state with new input, they add the two, while the forget gate still relies on multiplication.

To update the cell state, we have the *input gate*. The information from the previous hidden state and from the current input is passed through the *sigmoid function*, in order to obtain values between 0 and 1: the closer to 0 means to forget, and the closer to 1 means to remember. As you might have already noticed in Figure 5.4, LSTMs also have a *forget gate*, even if their specialty is actually to maintain and transfer information from distant occurrences to a final output. The forget gate decides what information should be forgotten and what information should be remembered and transferred.

Finally, we have the *output gate*, which decides what the next hidden state should be. First, the previous hidden state and the current input are passed into a sigmoid function. Then the newly modified cell state is passed through the tanh function. The tanh output is multiplied with the sigmoid output to decide what information the hidden state should keep (Zhang et al. 2019).

In the next section, we are going discuss a specific type of LSTM network, called gated recurrent units: they aim to solve the vanishing gradient problem that comes with a standard recurrent neural network. The gated recurrent unit, introduced by KyungHyun Cho and his colleagues in 2014 (Cho et al. 2014), is a slightly more streamlined variant of LSTM networks that often offers comparable performance and is significantly faster to compute.

Gated Recurrent Unit

In the previous section, we discussed how gradients are calculated in a recurrent neural network. In particular, you saw that long products of matrices can lead to vanishing or divergent gradients. GRU can also be considered as a variation on the LSTM because both are designed similarly and, in some cases, produce equally excellent results.

GRU networks do not leverage the cell state and use the hidden state to transfer information. Unlike LSTM, GRU networks contain only three gates and do not maintain an internal cell state. The information that is stored in the internal cell state in an LSTM recurrent unit is incorporated into the hidden state of the GRU. This combined information is aggregated and transferred to the next GRU (Cho et al. 2014).

The first two gates, *reset gate* and *update gate*, help solve the vanishing gradient problem of a standard RNN: these gates are two vectors that decide what information should be passed to the output (Cho et al. 2014). The unique characteristics about them is that they can be trained to keep information from long ago, without removing it through time, or delete information that is irrelevant to the prediction.

- *Update gate* – The update gate helps the model to determine how much of the past information (from previous time steps) needs to be passed along to the future. That is really powerful because the model can decide to copy all the information from the past and eliminate the risk of the vanishing gradient problem.

- *Reset gate* – The reset gate is another gate used to decide how much past information to forget.

Finally there is a third gate:

- *Current memory gate* – This gate is incorporated into the reset gate and is used to initiate some nonlinearity into the input and develop the input zero-mean. Another motivation to incorporate it into the reset gate is to decrease the impact that previous information has on the current information that is being passed into the future (Cho et al. 2014).

A GRU is a very useful mechanism for fixing the vanishing gradient problem in recurrent neural networks. The vanishing gradient dilemma occurs in machine learning when the gradient becomes small, which precludes the weight from adjusting its value.

In the next section, we are going discuss how you need to prepare your time series data for LSTMs and GRUs. Time series data requires preparation before it can be used to train a supervised learning model, such as an LSTM neural network. A time series must be transformed into samples with input and output components.

How to Prepare Time Series Data for LSTMs and GRUs

Your time series data set needs to be transformed before it can be utilized to fit a supervised learning model. As you learned in Chapter 3, Time Series Data Preparation, you first need to load your time series data in a pandas DataFrame so that the following operations can be performed:

1. Index the data on time stamp for time-based filtering

2. Visualize your data in tables with named columns

3. Identify missing periods and missing values

4. Create leading, lagged, and additional variables

Moreover, your time series data needs to be transformed in two tensors, as illustrated in Figure 5.5.

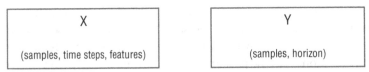

Figure 5.5: Transforming time series data into two tensors

As you can see from Figure 5.5, the number of samples, time steps, and features represents what is required to make a prediction (X), while sample and horizon represent what prediction is made (y). If we take the example of a univariate time series problem (for example, if we want to predict next energy load amount) where we are interested in one-step predictions (for example, next hour), the observations at prior time steps (for example, four energy load values during the four previous hours), so-called lag observations, are used as input and the output is the observation at the next time step (horizon), as illustrated Figure 5.6.

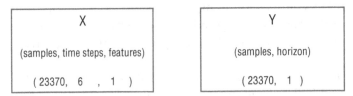

Figure 5.6: Transforming time series data into two tensors for a univariate time series problem

First of all, we need separate our dataset into train, validation, and test sets; in this way we can perform the following important steps:

1. We train the model on the train set.

2. The validation set is then used to evaluate the model after each training epoch and ensure that the model is not overfitting the training data.

3. After the model has finished training, we evaluate the model on the test set.

4. When handling time series data, it is important to ensure that both the validation set and test set cover a later period in time from the train set so that the model does not benefit from information from future time stamps.

Let's first import all necessary Python libraries and functions:

```
# Import necessary libraries
import os
import warnings
import matplotlib.pyplot as plt
import numpy as np
import pandas as pd
import datetime as dt
from collections import UserDict
from sklearn.preprocessing import MinMaxScaler
from IPython.display import Image
%matplotlib inline

from common.utils import load_data, mape

pd.options.display.float_format = '{:,.2f}'.format
np.set_printoptions(precision=2)
warnings.filterwarnings("ignore")
```

For our energy forecasting example, we allocate the period 1 November 2014 to 31 December 2014 to the test set. The period 1 September 2014 to 31 October is allocated to a validation set. All other time periods are available for the train set:

```
valid_st_data_load = '2014-09-01 00:00:00'
test_st_data_load = '2014-11-01 00:00:00'

ts_data_load[ts_data_load.index < valid_st_data_load][['load']].rename
(columns={'load':'train'}) \
    .join(ts_data_load[(ts_data_load.index >=valid_st_data_load)
& (ts_data_load.index < test_st_data_load)][['load']] \
        .rename(columns={'load':'validation'}), how='outer') \
    .join(ts_data_load[test_st_data_load:][['load']]
.rename(columns={'load':'test'}), how='outer') \
    .plot(y=['train', 'validation', 'test'], figsize=(15, 8),
fontsize=12)
plt.xlabel('timestamp', fontsize=12)
plt.ylabel('load', fontsize=12)
plt.show()
```

The code sample above will plot our train, validation, and test data sets, as illustrated in Figure 5.7.

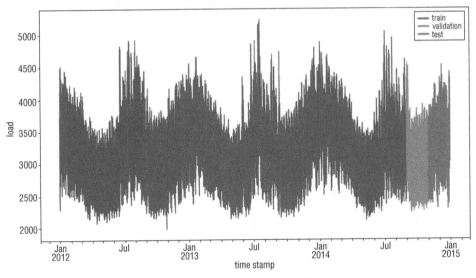

Figure 5.7: Ts_data_load train, validation, and test data sets plot

As shown in the sample code below, we are setting T (the number of lag variables) to 6. This means that the input for each sample is a vector of the previous 6 hours of the load values. The choice of T = 6 was arbitrary; you should consult business experts and experiment with different options before selecting one. We are also setting the horizon to 1, as we are interested in predicting next hour (t + 1) of our ts_data_load set:

```
T = 6
HORIZON = 1
```

Our data preparation for the train set will involve the following steps (Figure 5.8). In the sample code below, we explain how to perform the first four steps of this data preparation process:

```
# Step 1: get the train data from the correct data range
train = ts_data_load.copy()[ts_data_load.index < valid_st_data_load]
[['load']]

# Step 2: scale data to be in range (0, 1).
scaler = MinMaxScaler()
train['load'] = scaler.fit_transform(train)

# Step 3: shift the dataframe to create the input samples
train_shifted = train.copy()
train_shifted['y_t+1'] = train_shifted['load'].shift(-1, freq='H')
```

```
for t in range(1, T+1):
    train_shifted[str(T-t)] = train_shifted['load'].shift(T-t, freq='H')
y_col = 'y_t+1'
X_cols = ['load_t-5',
            'load_t-4',
            'load_t-3',
            'load_t-2',
            'load_t-1',
            'load_t']
train_shifted.columns = ['load_original']+[y_col]+X_cols

# Step 4: discard missing values
train_shifted = train_shifted.dropna(how='any')
train_shifted.head(5)
```

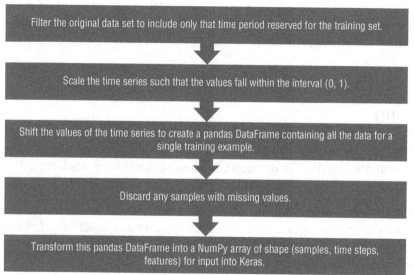

Figure 5.8: Data preparation steps for the ts_data_load train data set

Now we need to perform the last step in the data preparation process and convert the target and input features into NumPy arrays. X needs to be in the shape (samples, time steps, features). In our ts_data_load set, we have 23370 samples, 6 time steps, and 1 feature (load):

```
# Step 5: transform this pandas dataframe into a numpy array
y_train = train_shifted[y_col].as_matrix()
X_train = train_shifted[X_cols].as_matrix()
```

At this point we are ready to reshape the X input into a three-dimensional array, as shown in the following sample code:

```
X_train = X_train.reshape(X_train.shape[0], T, 1)
```

We now have a vector for target variable of shape:

`y_train.shape`

The tensor for the input features now has the shape:

`X_train.shape`

Finally, you need to follow the same process and steps listed above for the validation data set and keep T hours from the train set in order to construct initial features.

In this section, you discovered how to transform a time series data set into a three-dimensional structure ready for fitting an LSTM or GRU model. Specifically, you learned how to transform a time series data set into a two-dimensional supervised learning format and how to transform a two-dimensional time series data set into a three-dimensional structure suitable for LSTMs and GRUs.

In the next section, you will discover how to develop a suite of GRU models for time series forecasting problems.

How to Develop GRUs and LSTMs for Time Series Forecasting

Both LSTM and GRU models can be applied to time series forecasting. In this section, you will discover how to develop GRU models for time series forecasting scenarios. This section is divided into four parts:

- *Keras* – In this section you will get an overview of Keras capabilities for time series forecasting: Keras is an open-source neural-network library written in Python. It is capable of running on top of different deep learning tools, such as TensorFlow.

- *TensorFlow* – TensorFlow is an open-source software library for high-performance numerical computation. Its flexible architecture allows easy model building and deployment of computation across a variety of platforms.

- *Univariate models* – Univariate time series refers to a time series that consists of single (scalar) observations recorded sequentially over equal time increments. In this section, you will learn how to apply GRU models to univariate time series data.

- *Multivariate models* – A multivariate time series has more than one time-dependent variable. Each variable not only depends on its past values but also has some dependency on other variables. In this section, you will learn how to apply GRU models to multivariate time series data.

Keras

Keras is a Python wrapper library capable of running on top of different deep learning tools, such as TensorFlow. Most importantly, Keras is an API that supports both convolutional networks and recurrent networks and runs seamlessly on central processing units (CPUs) and graphics processing units (GPUs). Moreover, Keras is developed using four guiding philosophies (Nguyen et al. 2019):

- *User friendliness and minimalism* – Keras is an API designed with user experience in mind and offers consistent and simple APIs to build and train deep learning models.

- *Modularity* – While using Keras, data scientists need to look at their model development cycle as a modular process, in which independent components (such as neural layers, cost functions, optimizers, activation functions) can be easily leveraged and combined in different ways to create new models.

- *Easy extensibility* – Related to the previous principle, the third Keras's principle is about providing users with the capability of extending and adding new modules to existing ones, in order to improve the model development cycle.

- *Work with Python* – As mentioned previously, Keras is a Python wrapper library and its models are defined in Python code (`keras.io`).

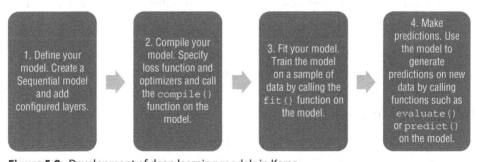

Figure 5.9: Development of deep learning models in Keras

As illustrated in Figure 5.9, the first step in the development process of deep learning models with Keras is the definition of a model. The main type of model is a sequence of layers called a *Sequential*, which is a linear stack of layers.

Once the model is defined, the second step is about compiling the model, which leverages the underlying framework to optimize the computation to be performed by your model. Once compiled, the model must be fit to data (third step in the development process of deep learning models with Keras). Finally, once it's trained, you can use your model to make predictions on new data.

In this section, you discovered the Keras Python library for deep learning research and development. In the next section, you will learn what TensorFlow is and why it's useful for data scientists when working with large-scale distributed training and inference.

TensorFlow

TensorFlow is an open-source library for numerical computation using data flow graphs. It was created and is maintained by the Google Brain team, and it is released under the Apache 2.0 open-source license. One of the most beneficial capabilities of TensorFlow is that data scientists can build and train their models with TensorFlow using the high-level Keras API, which makes getting started with TensorFlow and machine learning easy (`tensorflow.org`).

Nodes in the graph represent mathematical operations, while the graph edges represent the multidimensional data arrays (tensors) communicated between them. The distributed TensorFlow architecture contains distributed master and worker services with kernel implementations (Nguyen et al. 2019). TensorFlow was created for use in research, development, and production systems because of its capability of running on single CPU systems, GPUs, mobile devices, and large-scale distributed systems.

Moreover, TensorFlow gives you the flexibility and control with features like the Keras Functional API for easy prototyping and fast debugging, and it supports an ecosystem of powerful add-on libraries and models to experiment with, including Ragged Tensors, TensorFlow Probability, Tensor2Tensor, and BERT (`tensorflow.org`).

In this section, you discovered the Keras Python library for deep learning research and development. In the next section, you will learn how to implement a GRU model for a univariate time series forecasting scenario.

Univariate Models

LSTMs and GRUs can be used to model univariate time series forecasting problems. These are problems containing a single series of observations, and a model is required to learn from the series of past observations to predict the next value in the sequence.

In this example, we will implement a simple RNN forecasting model with the structure illustrated in Figure 5.10.

Let's now import all the necessary Keras packages, in order to define our model:

```
# Import necessary packages
from keras.models import Model, Sequential
from keras.layers import GRU, Dense
from keras.callbacks import EarlyStopping
```

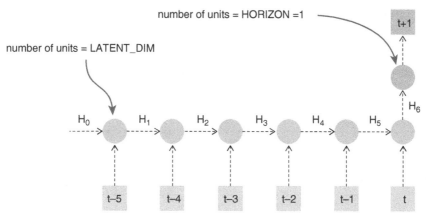

Figure 5.10: Structure of a simple RNN model to be implemented with Keras

There are a few aspects that need to be defined at this point:

- The latent dimension (LATENT_DIM): this is number of units in the RNN layer.

- The batch size (BATCH_SIZE): this is the number of samples per mini-batch.

- The epochs (EPOCHS): this is the maximum number of times the training algorithm will cycle through all samples.

Let's define them in the sample code below:

```
LATENT_DIM = 5
BATCH_SIZE = 32
EPOCHS = 10
```

Now we can define our model and create a Sequential model, as the sample code below shows:

```
model = Sequential()
model.add(GRU(LATENT_DIM, input_shape=(T, 1)))
model.add(Dense(HORIZON))
```

As the next step, we need to compile our model. We specify the loss function and optimizers and call the compile() function on the model. For this specific example, we use the mean squared error as the loss function. The Keras documentation recommends the optimizer RMSprop for RNNs:

```
model.compile(optimizer='RMSprop', loss='mse')
model.summary()
```

Running the sample code above will output the following summary table:

```
Layer (type)                    Output Shape              Param #
=================================================================
gru_1 (GRU)                     (None, 5)                 105

dense_1 (Dense)                 (None, 1)                 6
=================================================================
Total params: 111
Trainable params: 111
Non-trainable params: 0
```

We now need to specify the early stopping criteria. Early stopping is a method of regularization used to avoid overfitting when training a machine learning model with an iterative method, such as gradient descent. In our example, we will use early stopping as a form of validation to detect when overfitting starts during supervised training of a neural network; training is stopped before convergence to avoid the overfitting ("early stopping").

In our case, we monitor the validation loss (in this case the mean squared error) on the validation set after each training epoch. If the validation loss has not improved by `min _ delta` after patience epochs, we stop the training, as shown in the sample code below:

```
GRU_earlystop = EarlyStopping(monitor='val_loss', min_delta=0,
patience=5
```

Now we can fit our model, by training it on a sample of data by calling the `fit()` function on the model itself:

```
model_history = model.fit(X_train,
                y_train,
                batch_size=BATCH_SIZE,
                epochs=EPOCHS,
                validation_data=(X_valid, y_valid),
                callbacks=[GRU_earlystop],
                verbose=1)
```

Let's now make predictions by using the model created to generate predictions on new data and by calling functions such as `evaluate()` or `predict()` on the model. In order to evaluate the model, you first need to perform the data preparation steps listed above on the test data set.

After the test data preparation steps, we are now ready to make the predictions on the test data set and compare those predictions to our actual load values:

```
ts_predictions = model.predict(X_test)
ts_predictions
```

```
ev_ts_data = pd.DataFrame(ts_predictions, columns=['t+'+str(t) for t in
range(1, HORIZON+1)])
ev_ts_data['timestamp'] = test_shifted.index
ev_ts_data = pd.melt(ev_ts_data, id_vars='timestamp',
value_name='prediction', var_name='h')
ev_ts_data['actual'] = np.transpose(y_test).ravel()
ev_ts_data[['prediction', 'actual']] = scaler.inverse_transform(ev_ts_
data[['prediction', 'actual']])
ev_ts_data.head()
```

Running the example above will produce the following results:

```
        timestamp                h        prediction    actual
0       2014-11-01 05:00:00      t+1      2,673.13      2,714.00
1       2014-11-01 06:00:00      t+1      2,947.12      2,970.00
2       2014-11-01 07:00:00      t+1      3,208.74      3,189.00
3       2014-11-01 08:00:00      t+1      3,337.19      3,356.00
4       2014-11-01 09:00:00      t+1      3,466.88      3,436.00
```

In order to evaluate our model, we can compute the mean absolute percentage error (MAPE) over all predictions, as demonstrated in the sample code below:

```
# %load -s mape common/utils.py
def mape(ts_predictions, actuals):
    """Mean absolute percentage error"""
    return ((ts_predictions - actuals).abs() / actuals).mean()


mape(ev_ts_data['prediction'], ev_ts_data['actual'])
```

The MAPE is 0.015, meaning that our model is 99.985 percent accurate. We can complete our exercise, by plotting the predictions versus the actuals for the first week of the test set:

```
ev_ts_data[ev_ts_data.timestamp<'2014-11-08']
.plot(x='timestamp', y=['prediction', 'actual'],
  style=['r', 'b'], figsize=(15, 8))
plt.xlabel('timestamp', fontsize=12)
plt.ylabel('load', fontsize=12)
plt.show()
```

Running the sample code above will plot the visualization illustrated in Figure 5.11.

In Figure 5.11, we can observe the predictions originated by our RNN model versus actual values.

Now that we have looked at GRU models for univariate data, let's turn our attention to multivariate data: when the data involves three or more variables, it is categorized under multivariate, as it contains more than one dependent variable. In the next section, we will discuss multivariate GRU models.

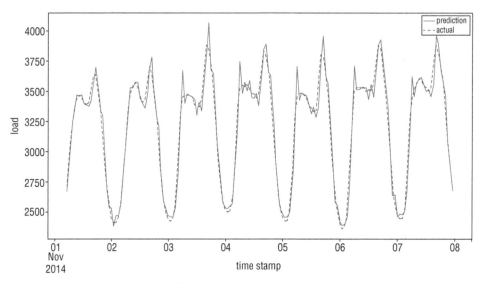

Figure 5.11: Structure of a simple RNN model to be implemented with Keras

Multivariate Models

Multivariate time series data means data where there is more than one observation for each time step. In this section, we demonstrate how to complete the following steps:

- Prepare time series data for training an RNN forecasting model.
- Get data in the required shape for the Keras API.
- Implement an RNN model in Keras to predict the next step ahead (time t + 1) in the time series. This model uses recent values of temperature, as well as load, as the model input.
- Enable early stopping to reduce the likelihood of model overfitting.
- Evaluate the model on a test data set.

For each step above, we will use our ts_data set as an example. Let's start by loading the data into a pandas DataFrame:

```
ts_data = load_data(data_dir)
ts_data.head()
```

Running the sample code above will output the following table:

	load	temp
2012-01-01 00:00:00	2,698.00	32.00
2012-01-01 01:00:00	2,558.00	32.67
2012-01-01 02:00:00	2,444.00	30.00
2012-01-01 03:00:00	2,402.00	31.00
2012-01-01 04:00:00	2,403.00	32.00

As you notice from the table above, we are now dealing with a multivariate data set, as we have two variables, load and temp. As we did in the previous section, we now need to create validation and test data sets, and define our T (number of lags) and our HORIZON (how many hours in the future we want to predict), as shown in the following sample code:

```
valid_st_data_load = '2014-09-01 00:00:00'
test_st_data_load = '2014-11-01 00:00:00'

T = 6
HORIZON = 1
```

We create the training data set with load and temp features, and we fit a scaler for the y values, as demonstrated in this sample code:

```
from sklearn.preprocessing import MinMaxScaler
y_scaler = MinMaxScaler()
y_scaler.fit(train[['load']])
```

We need to also scale the input features data (the load and temp values):

```
X_scaler = MinMaxScaler()
train[['load', 'temp']] = X_scaler.fit_transform(train)
```

In our example, we are going to use the `TimeSeriesTensor` convenience class to perform the following steps:

1. Shift the values of the time series to create a pandas DataFrame containing all the data for a single training example.
2. Discard any samples with missing values.
3. Transform this pandas DataFrame into a NumPy array of shape (samples, time steps, and features) for input into Keras.

This class takes the following parameters, as also demonstrated in the sample code below:

- ▪ `dataset`: original time series
- ▪ `H`: the forecast horizon
- ▪ `tensor _ structure`: a dictionary describing the tensor structure in the form `{ 'tensor_name' : (range(max_backward_shift, max_forward_ shift), [feature, feature, ...] ) }`
- ▪ `freq`: time series frequency
- ▪ `drop_incomplete`: (Boolean) whether to drop incomplete samples

```
tensor = {'X':(range(-T+1, 1), ['load', 'temp'])}
ts_train_inp = TimeSeriesTensor(dataset=train,
```

Continues

(*continued*)

```
target='load',
H=HORIZON,
tensor_structure=tensor,
freq='H',
drop_incomplete=True)
```

Finally, we need to define our validation data set, as shown in the sample code below:

```
back_ts_data = dt.datetime.strptime
(valid_st_data_load, '%Y-%m-%d %H:%M:%S')
- dt.timedelta(hours=T-1)
ts_data_valid = ts_data.copy()
[(ts_data.index >=back_ts_data)
& (ts_data.index < test_st_data_load)][['load', 'temp']]
ts_data_valid[['load', 'temp']] = X_scaler.transform(ts_data_valid)
ts_data_valid_inputs = TimeSeriesTensor(ts_data_valid, 'load', HORIZON,
tensor)
```

We are now ready to implement a simple RNN forecasting model for our multivariate forecasting scenario with the following structure (Figure 5.12).

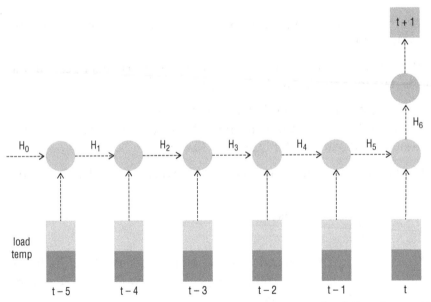

Figure 5.12: Structure of a simple RNN model to be implemented with Keras for a multivariate forecasting scenario

In the following sample code, we will import the necessary Keras packages; we will define the latent dimension, batch size, and epochs for our model; and we will define the model itself:

```
# Import necessary packages
from keras.models import Model, Sequential
from keras.layers import GRU, Dense
from keras.callbacks import EarlyStopping

# Define parameters
LATENT_DIM = 5
BATCH_SIZE = 32
EPOCHS = 50

#Define model
model = Sequential()
model.add(GRU(LATENT_DIM, input_shape=(T, 2)))
model.add(Dense(HORIZON))
```

We can now compile our model, by specifying the loss function and optimizers and calling the `compile()` function on the model, and have a look at its summary:

```
model.compile(optimizer='RMSprop', loss='mse')

model.summary()
```

Running the sample code above will generate the following summary table:

Layer (type)	Output Shape	Param #
gru_1 (GRU)	(None, 5)	120
dense_1 (Dense)	(None, 1)	6

```
Total params: 126
Trainable params: 126
Non-trainable params: 0
```

We can now fit our multivariate model and train it on a sample of data by calling the `fit()` function on the model:

```
GRU_earlystop = EarlyStopping(monitor='val_loss', min_delta=0,
patience=5)
```

Continues

(*continued*)

```
model_history = model.fit(ts_train_inp['X'],
                    ts_train_inp['target'],
                    batch_size=BATCH_SIZE,
                    epochs=EPOCHS,
                     validation_data=(valid_inputs['X'],
        valid_inputs['target']),
                    callbacks=[GRU_earlystop],
                    verbose=1)
```

Once we have defined the model, we can use it to generate predictions on new data (in this example we will use our test data set) by calling functions such as evaluate():

```
back_ts_data = dt.datetime.strptime
(test_st_data_load, '%Y-%m-%d %H:%M:%S')
- dt.timedelta(hours=T-1)
ts_data_test = ts_data.copy()[test_st_data_load:][['load', 'temp']]
ts_data_test[['load', 'temp']] = X_scaler.transform(ts_data_test)
test_inputs = TimeSeriesTensor(ts_data_test, 'load', HORIZON, tensor)

ts_predictions = model.predict(test_inputs['X'])

ev_ts_data = create_evaluation_df(ts_predictions, test_inputs, HORIZON,
y_scaler)
ev_ts_data.head()
```

Running the sample code above will generate the following table with predictions and actual values:

	Timestamp	h	prediction	actual
0	2014-11-01 05:00:00	t+1	2,737.52	2,714.00
1	2014-11-01 06:00:00	t+1	2,975.03	2,970.00
2	2014-11-01 07:00:00	t+1	3,204.60	3,189.00
3	2014-11-01 08:00:00	t+1	3,326.30	3,356.00
4	2014-11-01 09:00:00	t+1	3,493.52	3,436.00

In order to visualize the overall accuracy of our model, we can estimate the MAPE, as demonstrated in the sample code below:

```
mape(ev_ts_data['prediction'], ev_ts_data['actual'])
```

The MAPE is 0.015, showing that our model is 99.985 percent accurate. It is important to note that, even if we decided to add a variable to our model to make it multivariate, the model performance did not improve.

Conclusion

Deep learning neural networks are powerful engines capable of learning from arbitrary mappings from inputs to outputs, supporting multiple inputs and outputs, and automatically extracting patterns in input data that spans over long sequences of time. All these characteristics together make neural networks helpful tools when dealing with more complex time series forecasting problems that involve large amounts of data, multiple variables with complicated relationships, and even multi-step time series tasks.

In this chapter we discussed some of the practical reasons data scientists may still want to think about deep learning when they build time series forecasting solutions.

Specifically, we took a closer look at the following important topics:

- *Reasons to Add Deep Learning to Your Time Series Toolkit* – You learned how deep learning neural networks are able to automatically learn arbitrary complex mappings from inputs to outputs and support multiple inputs and outputs.

- *Recurrent Neural Networks for Time Series Forecasting* – In this section, I introduced a very popular type of artificial neural network: recurrent neural networks, also known as RNNs. We also discussed a variation of recurrent neural networks, the so-called long short-term memory units, and you learned how to prepare time series data for LSTMs and GRU models.

- *How to Develop GRUs and LSTMs for Time Series Forecasting* – In this final section of Chapter 5, you learned how to develop long short-term memory models for time series forecasting. In particular, you learned how to develop GRU models for time series forecasting problems.

In the next chapter, Chapter 6, "Model Deployment for Time Series Forecasting," you will learn how to deploy your machine learning models as web services: the purpose of the next chapter is to provide a complete overview of tools and concepts that you need in order to operationalize and push into production your time series forecasting solutions.

Model Deployment for Time Series Forecasting

Throughout the book, I introduced a few real-world data science scenarios that I used to showcase some of the key time series concepts, steps, and codes. In this final chapter, I will walk you through the process of building and deploying some of the time series forecasting solutions by employing some of these use cases and data sets.

The purpose of this chapter is to provide a complete overview of tools to build and deploy your own time series forecasting solutions by discussing the following topics:

> ➤ *Experimental Set Up and Introduction to Azure Machine Learning SDK for Python* – In this section, I will introduce Azure Machine Learning SDK for Python to build and run machine learning workflows. You will get an overview of some of the most important classes in the SDK and how you can use them to build, train, and deploy a machine learning model on Azure.

Specifically, in this section you will discover the following concepts and assets:

- ■ `Workspace`, which is a foundational resource in the cloud that you use to experiment, train, and deploy machine learning models.

- ■ `Experiment`, which is another foundational cloud resource that represents a collection of trials (individual model runs).

- ▪ `Run`, which represents a single trial of an experiment.

- ▪ `Model`, which is used for working with cloud representations of the machine learning model.

- ▪ `ComputeTarget, RunConfiguration, ScriptRunConfig`, which are abstract parent classes for creating and managing compute targets. A compute target represents a variety of resources you can use to train your machine learning models.

- ▪ `Image`, which is an abstract parent class for packaging models into container images that include the runtime environment and dependencies.

- ▪ `Webservice`, which is the abstract parent class for creating and deploying web services for your models.

- ▪ *Machine Learning Model Deployment* – In this section, we will talk more about machine learning model deployment, that is, the method of integrating a machine learning model into an existing production environment in order to begin developing practical business decisions based on data. Through machine learning model deployment, companies can begin to take full advantage of the predictive and intelligent models they build and, therefore, transform themselves into actual AI-driven businesses.

- ▪ *Solution Architecture for Time Series Forecasting with Deployment Examples* - In this final section of the chapter, we will build, train, and deploy a demand forecasting solution. I will demonstrate how to build an end-to-end data pipeline architecture and deployment code that can be generalized for different time series forecasting solutions.

Experimental Set Up and Introduction to Azure Machine Learning SDK for Python

Azure Machine Learning provides SDKs and services for data scientists and developers to prepare data and train and deploy machine learning models. In this chapter, we will use Azure Machine Learning SDK for Python (`aka.ms/AzureMLSDK`) to build and run machine learning workflows.

The following sections are a summary of some of the most important classes in the SDK that you can use to build your time series forecasting solution: you can find all information about the classes below on the official website of Azure Machine Learning SDK for Python.

Workspace

The `Workspace` is a Python-based function that you can use to experiment, train, and deploy machine learning models. You can import the class and create a new workspace by using the following code:

```
from azureml.core import Workspace
ws = Workspace.create(name='myworkspace',
                      subscription_id='<your-azure-subscription-id>',
                      resource_group='myresourcegroup',
                      create_resource_group=True,
                      location='eastus2'
                      )
```

It is recommended that you set `create_resource_group` to False if you have a previously existing Azure resource group that you want to use for the workspace. Some functions might prompt for Azure authentication credentials. For more information on the `Workspace` class in Azure ML SDK for Python, visit aka.ms/AzureMLSDK.

Experiment

The `Experiment` is a cloud resource that embodies a collection of trials (individual model runs). The following code fetches an experiment object from within `Workspace` by name, or it creates a new experiment object if the name does not exist (aka.ms/AzureMLSDK):

```
from azureml.core.experiment import Experiment
experiment = Experiment(workspace=ws, name='test-experiment')
```

You can run the following code to get a list of all experiment objects contained in `Workspace`, as shown in the code below:

```
list_experiments = Experiment.list(ws)
```

For more information on the `Experiment` class in Azure ML SDK for Python, visit aka.ms/AzureMLSDK.

Run

The `Run` class represents a single trial of an experiment. A `Run` is an object that you use to monitor the asynchronous execution of a trial, store the output of the trial, analyze results, and access generated artifacts. You can use a `Run` inside your experimentation code to log metrics and artifacts to the `Run History` service (aka.ms/AzureMLSDK).

In the following code, I show how to create a run object by submitting an experiment object with a run configuration object:

```
tags = {"prod": "phase-1-model-tests"}
run = experiment.submit(config=your_config_object, tags=tags)
```

As you notice, you can use the `tags` parameter to attach custom categories and labels to your runs. Moreover, you can use the static list function to get a list of all run objects from an experiment. You need to specify the `tags` parameter to filter by your previously created tag:

```
from azureml.core.run import Run
filtered_list_runs = Run.list(experiment, tags=tags)
```

For more information on the `Run` class in Azure ML SDK for Python, visit aka.ms/AzureMLSDK.

Model

The `Model` class is used for working with cloud representations of different machine learning models. You can use model registration to store and version your models in your workspace in the Azure cloud. Registered models are identified by name and version. Each time you register a model with the same name as an existing one, the registry increments the version (aka.ms/AzureMLSDK).

The following example shows how to build a simple local classification model with scikit-learn, register the model in the workspace, and download the model from the cloud:

```
from sklearn import svm
import joblib
import numpy as np

# customer ages
X_train = np.array([50, 17, 35, 23, 28, 40, 31, 29, 19, 62])
X_train = X_train.reshape(-1, 1)
# churn y/n
y_train = ["yes", "no", "no", "no", "yes", "yes", "yes", "no", "no", "yes"]

clf = svm.SVC(gamma=0.001, C=100.)
clf.fit(X_train, y_train)

joblib.dump(value=clf, filename="churn-model.pkl")
```

Moreover, you can use the `register` function to register the model in your workspace:

```
from azureml.core.model import Model
model = Model.register(workspace=ws,
model_path="churn-model.pkl",
model_name="churn-model-test")
```

After you have a registered model, deploying it as a web service is a simple process:

1. You need to create and register an image. This step configures the Python environment and its dependencies.

2. As the second step, you need create an image.

3. Finally, you need to attach your image.

For more information on the `Run` class in Azure ML SDK for Python, visit aka.ms/AzureMLSDK.

Compute Target, RunConfiguration, and ScriptRunConfig

The `ComputeTarget` class is a parent class for creating and managing compute targets. A compute target represents a variety of resources where you can train your machine learning models. A compute target can be either a local machine or a cloud resource, such as Azure Machine Learning Compute, Azure HDInsight, or a remote virtual machine (aka.ms/AzureMLSDK).

First of all, you need to set up an `AmlCompute` (child class of `ComputeTarget`) target. For the sample below, we can reuse the simple scikit-learn churn model and build it into its own file, `train.py`, in the current directory (aka.ms/AzureMLSDK). At the end of the file, we create a new directory called `outputs` to store your trained model that `joblib.dump()` serialized:

```
# train.py
from sklearn import svm
import numpy as np
import joblib
import os

# customer ages
X_train = np.array([50, 17, 35, 23, 28, 40, 31, 29, 19, 62])
X_train = X_train.reshape(-1, 1)
# churn y/n
y_train = ["cat", "dog", "dog", "dog", "cat", "cat", "cat", "dog",
"dog", "cat"]
clf = svm.SVC(gamma=0.001, C=100.)
clf.fit(X_train, y_train)
```

Continues

continued
```
os.makedirs("outputs", exist_ok=True)
joblib.dump(value=clf, filename="outputs/churn-model.pkl")
```

Next, you can create the compute target by instantiating a `RunConfiguration` object and setting the type and size (aka.ms/AzureMLSDK):

```
from azureml.core.runconfig import RunConfiguration
from azureml.core.compute import AmlCompute
list_vms = AmlCompute.supported_vmsizes(workspace=ws)
compute_config = RunConfiguration()
compute_config.target = "amlcompute"
compute_config.amlcompute.vm_size = "STANDARD_D1_V2"
```

Now you are ready to submit the experiment by using the `ScriptRunConfig` and specifying the `config` parameter of the `submit()` function:

```
from azureml.core.experiment import Experiment
from azureml.core import ScriptRunConfig
script_run_config = ScriptRunConfig(source_directory=os.getcwd(),
script="train.py", run_config=compute_config)
experiment = Experiment(workspace=ws, name="compute_target_test")
run = experiment.submit(config=script_run_config)
```

For more information on these classes in Azure ML SDK for Python, visit `aka.ms/AzureMLSDK`.

Image and Webservice

The `Image` class is a parent class for packaging models into container images that include the runtime environment and dependencies. The `Webservice` class is another parent class for creating and deploying web services for your models (`aka.ms/AzureMLSDK`).

The following code shows a basic example of creating an image and using it to deploy a web service. The `ContainerImage` class extends an image and creates a Docker image.

```
from azureml.core.image import ContainerImage
image_config = ContainerImage.image_configuration(execution_script="score.
py",
                                                  runtime="python",
                                                  conda_file="myenv.yml",
                                        description="test-image-config")
```

In this example, `score.py` processes the request/response for the web service. The script defines two methods: `init()` and `run()`.

```
image = ContainerImage.create(name="test-image",
                              models=[model],
```

```
                                    image_config=image_config,
                                    workspace=ws)
```

To deploy the image as a web service, you first need to build a deployment configuration, as shown in the following sample code:

```
from azureml.core.webservice import AciWebservice
deploy_config = AciWebservice.deploy_configuration(cpu_cores=1,
memory_gb=1)
```

After, you can use the deployment configuration to create a web service, as shown in the sample code below:

```
from azureml.core.webservice import Webservice
service = Webservice.deploy_from_image(deployment_config=deploy_config,
                                       image=image,
                                       name=service_name,
                                       workspace=ws
                                       )
service.wait_for_deployment(show_output=True)
```

In this first part of this chapter, I introduced some of the most important classes in the SDK (for more information, visit aka.ms/AzureMLSDK) and common design patterns for using them. In the next section, we will look at the machine learning deployment on Azure Machine Learning.

Machine Learning Model Deployment

Model deployment is the method of integrating a machine learning model into an existing production environment in order to begin developing practical business decisions based on data. It is only once models are deployed to production that they start adding value, making deployment a crucial step (Lazzeri 2019c).

Model deployment is a fundamental step of the machine learning model workflow (Figure 6.1). Through machine learning model deployment, companies can begin to take full advantage of the predictive and intelligent models they build and, therefore, transform themselves into actual data-driven businesses.

When we think about machine learning, we focus our attention on key components such as data sources, data pipelines, how to test machine learning models at the core of our machine learning application, how to engineer our features, and which variables to use to make the models more accurate. All these steps are important; however, thinking about how we are going to consume those models and data over time is also a critical step in the machine learning pipeline. We can only begin extracting real value and business benefits from a model's predictions when it has been deployed and operationalized.

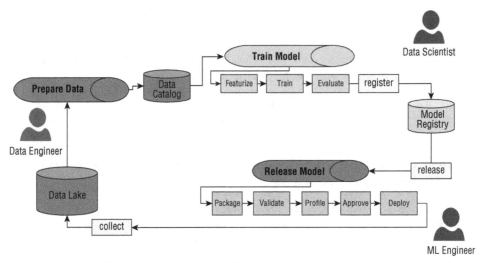

Figure 6.1: The machine learning model workflow

Successful model deployment is fundamental for data-driven enterprises for the following key reasons:

- Deployment of machine learning models means making models available to external customers and/or other teams and stakeholders in your company.

- When you deploy models, other teams in your company can use them, send data to them, and get their predictions, which are in turn populated back into the company systems to increase training data quality and quantity.

- Once this process is initiated, companies will start building and deploying higher numbers of machine learning models in production and master robust and repeatable ways to move models from development environments into business operations systems (Lazzeri 2019c).

From an organizational perspective, many companies see AI-enablement as a technical effort. However, it is more of a business-driven initiative that starts within the company; in order to become an AI-driven company, it is important that the people who successfully operate and understand the business today are also the ones who are responsible for building and driving the machine learning pipeline, from model training to model deployment and monitoring.

Right from the first day of a machine learning process, machine learning teams should interact with business partners. It is essential to maintain constant interaction to understand the model experimentation process parallel to the model deployment and consumption steps. Most organizations struggle to unlock machine learning's potential to optimize their operational processes and get data scientists, analysts, and business teams speaking the same language (Lazzeri 2019c).

Moreover, machine learning models must be trained on historical data, which demands the creation of a prediction data pipeline, an activity requiring multiple tasks including data processing, feature engineering, and tuning. Each task, down to versions of libraries and handling of missing values, must be exactly duplicated from the development to the production environment. Sometimes, differences in technology used in development and in production contribute to difficulties in deploying machine learning models.

Companies can use machine learning pipelines to create and manage workflows that stitch together machine learning phases. For example, a pipeline might include data preparation, model training, model deployment, and inference/scoring phases. Each phase can encompass multiple steps, each of which can run unattended in various compute targets.

How to Select the Right Tools to Succeed with Model Deployment

Current approaches of handcrafting machine learning models are too slow and unproductive for companies intent on transforming their operations with AI. Even after months of development, which delivers a model based on a single algorithm, the management team has little means of knowing whether their data scientists have created a great model and how to operationalize it (Lazzeri 2019c).

Below, I share a few guidelines on how a company can select the right tools to succeed with model deployment. I will illustrate this workflow using Azure Machine Learning Service, but it can be also used with any machine learning product of your choice.

The model deployment workflow should be based on the following three simple steps:

- Register the model.
- Prepare to deploy (specify assets, usage, compute target).
- Deploy the model to the compute target.

As we saw in the previous section, Model is the logical container for one or more files that make up your model. For example, if you have a model that is stored in multiple files, you can register them as a single model in the workspace. After registration, you can then download or deploy the registered model and receive all the files that were registered.

Machine learning models are registered when you create an Azure Machine Learning workspace. The model can come from Azure Machine Learning or it can come from somewhere else.

To deploy your model as a web service, you must create an inference configuration (InferenceConfig) and a deployment configuration. Inference, or model scoring, is the phase where the deployed model is used for prediction, most commonly on production data. In the `inferenceConfig`, you specify the scripts and dependencies needed to serve your model. In the deployment configuration you specify details of how to serve the model on the compute target (Lazzeri 2019c).

The entry script receives data submitted to a deployed web service and passes it to the model. It then takes the response returned by the model and returns that to the client. The script is specific to your model; it must understand the data that the model expects and returns.

The script contains two functions that load and run the model:

- `init()` – Typically, this function loads the model into a global object. This function is run only once when the Docker container for your web service is started.

- `run(input_data)` – This function uses the model to predict a value based on the input data. Inputs and outputs to the run typically use JSON for serialization and de-serialization. You can also work with raw binary data. You can transform the data before sending to the model or before returning to the client.

When you register a model, you provide a model name used for managing the model in the registry. You use this name with the `Model.get_model_path()` to retrieve the path of the model file(s) on the local file system. If you register a folder or a collection of files, this API returns the path to the directory that contains those files.

Finally, before deploying, you must define the deployment configuration. The deployment configuration is specific to the compute target that will host the web service. For example, when deploying locally, you must specify the port where the service accepts requests. The following compute targets, or compute resources, can be used to host your web service deployment:

- *Local web service and notebook virtual machine (VM) web service* – Both compute targets are used for testing and debugging. They are considered good for limited testing and troubleshooting.

- *Azure Kubernetes Service (AKS)* – This compute target is used for real-time inference. It is considered good for high-scale production deployments.

- *Azure Container Instances (ACI)* – This compute target is used for testing. It is considered good for low-scale, CPU-based workloads requiring <48 GB RAM.

- *Azure Machine Learning Compute* – This compute target is used for batch inference as it is able to run batch scoring on serverless compute targets.

- *Azure IoT Edge* – This is an IoT module, able to deploy and serve machine learning models on IoT devices.

- *Azure Stack Edge* – Developers and data scientists can use this compute target via IoT Edge.

In this section, I introduced some common challenges of machine learning model deployment, and we discussed why successful model deployment is fundamental to unlock the full potential of AI, why companies struggle with model deployment, and how to select the right tools to succeed with model deployment.

Next, we will apply what you learned in the first two sections of this chapter to a real demand forecasting use case.

Solution Architecture for Time Series Forecasting with Deployment Examples

In this final section of this chapter, we will build, train, and deploy an energy demand forecasting solution on Azure. For this specific use case, we will use data from the GEFCom2014 energy forecasting competition. For more information, please refer to "Probabilistic Energy Forecasting: Global Energy Forecasting Competition 2014 and Beyond" (Tao Hong et al. 2016).

The raw data consists of rows and columns. Each measurement is represented as a single row of data. Each row of data includes multiple columns (also referred to as features or fields). After identifying the required data sources, we would like to ensure that the raw data that has been collected includes the correct data features. To build a reliable demand forecast model, we would need to ensure that the data collected includes data elements that can help predict future demand. Here are some basic requirements concerning the data structure (schema) of the raw data:

- *Time stamp* – The time stamp field represents the actual time the measurement was recorded. It should comply with one of the common date/time formats. Both date and time parts should be included. In most cases there is no need for the time to be recorded till the second level of granularity. It is important to specify the time zone in which the data is recorded.

- *Load* – Hourly historical load data for the utility were provided. This is the actual consumption at a given date/time. The consumption can be measured in kWh (kilowatt-hour) or any other preferred unit. It is important to note that the measurement unit must stay consistent across all measurements in the data. In some cases, consumption can be supplied over three power phases. In that case we would need to collect all the independent consumption phases.

▪ *Temperature* – Hourly historical temperature data for the utility were provided. The temperature is typically collected from an independent source. However, it should be compatible with the consumption data. It should include a time stamp as described above that will allow it to be synchronized with the actual consumption data. The temperature value can be specified in degrees Celsius or Fahrenheit but should stay consistent across all measurements.

The modeling phase is where the conversion of the data into a model takes place. In the core of this process there are advanced algorithms that scan the historical data (training data), extract patterns, and build a model. That model can be later used to predict on new data that has not been used to build the model.

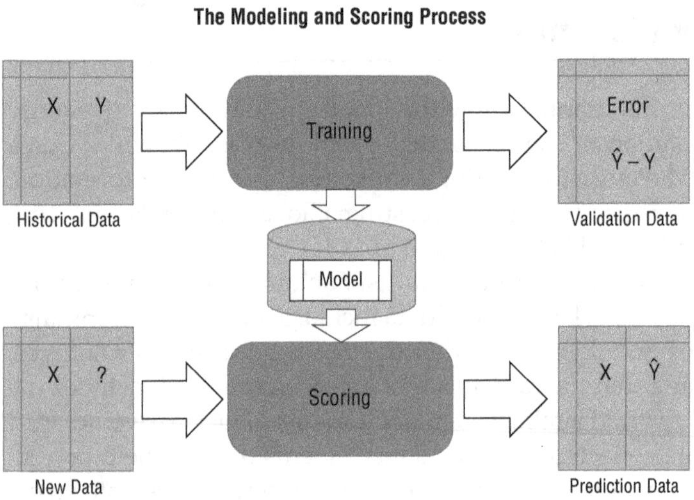

The Modeling and Scoring Process

Figure 6.2: The modeling and scoring process

As can be seen from Figure 6.2, the historical data feeds the training module. Historical data is structured where the independent feature is denoted as X and the dependent (target) variable is denoted as Y. Both X and Y are produced during the data preparation process. The training module consists of an algorithm that scans the data and learns its features and patterns. The actual algorithm is selected by the data scientist and should best match the type of the problem we attempt to solve.

The training algorithms are usually categorized as regression (predict numeric outcomes), classification (predict categorical outcomes), clustering (identify groups), and forecasting. The training module generates the model as an object that can be stored for future use. During training, we can also quantify the prediction accuracy of the model by using validation data and measuring the prediction error.

Once we have a working model, we can then use it to score new data that is structured to include the required features (X). The scoring process will make use of the persisted model (object from the training phase) and predict the target variable that is denoted by $\hat{Y}$.

In case of demand forecasting, we make use of historical data that is ordered by time. We generally refer to data that includes the time dimension as time series. The goal in time series modeling is to find time-related trends, seasonality, and autocorrelation (correlation over time) and formulate those into a model. In recent years, advanced algorithms have been developed to accommodate time series forecasting and to improve forecasting accuracy.

Train and Deploy an ARIMA Model

In the next few sections, I will show how to build, train, and deploy an ARIMA model for energy demand forecasting. Let's start with the data setup: the data in this example is taken from the GEFCom2014 forecasting competition. It consists of three years of hourly electricity load and temperature values between 2012 and 2014.

Let's import the necessary Python modules to get started:

```
# Import modules
import os
import shutil
import matplotlib.pyplot as plt
from common.utils import load_data, extract_data, download_file
%matplotlib inline
```

As a second step, you need to download the data and store it in a data folder:

```
data_dir = './data'

if not os.path.exists(data_dir):
    os.mkdir(data_dir)

if not os.path.exists(os.path.join(data_dir, 'energy.csv')):
    # Download and move the zip file
    download_file("https://mlftsfwp.blob.core.windows.net/mlftsfwp/GEFCom2014.zip")
    shutil.move("GEFCom2014.zip", os.path.join(data_dir,"GEFCom2014.zip"))
    # If not done already, extract zipped data and save as csv
    extract_data(data_dir)
```

Once you have completed the task above, you are ready to load the data from CSV into a pandas DataFrame:

```
energy = load_data(data_dir)[['load']]
energy.head()
```

This code will produce the output illustrated in Figure 6.3.

	load
2012-01-01 00:00:00	2698.0
2012-01-01 01:00:00	2558.0
2012-01-01 02:00:00	2444.0
2012-01-01 03:00:00	2402.0
2012-01-01 04:00:00	2403.0

Figure 6.3: First few rows of the energy data set

In order to visualize our data set and make sure that all data was uploaded, let's first plot all available load data (January 2012 to December 2014):

```
energy.plot(y='load', subplots=True, figsize=(15, 8), fontsize=12)
plt.xlabel('timestamp', fontsize=12)
plt.ylabel('load', fontsize=12)
plt.show()
```

The code above will output the plot shown in Figure 6.4.

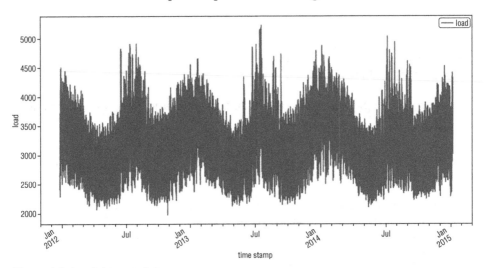

Figure 6.4: Load data set plot

In the preceding example (Figure 6.4), we plot the first column of our data set (the time stamp is taken as an index of the DataFrame). If you want to print another column, the variable `column_to_plot` can be adjusted.

Now let's visualize a subsample of the data, by plotting the first week of July 2014:

```
energy['7/1/2014':'7/7/2014'].plot
(y=column_to_plot, subplots=True,
figsize=(15, 8), fontsize=12)
plt.xlabel('timestamp', fontsize=12)
plt.ylabel(column_to_plot, fontsize=12)
plt.show(
```

This will create the plot with the data points from the first week of July 2014, as shown in Figure 6.5.

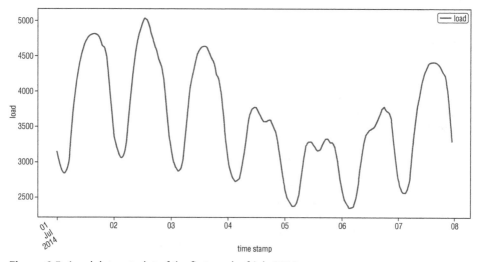

Figure 6.5: Load data set plot of the first week of July 2014

If you are able to run this notebook successfully and see all the visualizations, you are ready to move to the training step. Let's start with the configuration part: At this point you need to set up your Azure Machine Learning services workspace and configure your notebook library. For more information, visit aka.ms/AzureMLConfiguration and follow the instructions in the notebook.

The training script executes a training experiment. Once the data is prepared, you can train a model and see the results on Azure.

There are several steps to follow:

- Configure the workspace.
- Create an experiment.
- Create or attach a compute cluster.
- Upload the data to Azure.
- Create an estimator.
- Submit the work to the remote cluster.
- Register the model.
- Deploy the model.

Let's start with importing the Azure Machine Learning Python SDK library and other modules and configuring the workspace.

Configure the Workspace

First of all, you need to import Azure Machine Learning Python SDK and other Python modules that you will need for the training script:

```
import datetime as dt
import math
import os
import urllib.request
import warnings

import azureml.core
import azureml.dataprep as dprep
import matplotlib.pyplot as plt
import numpy as np
import pandas as pd
from azureml.core import Experiment, Workspace
from azureml.core.compute import AmlCompute, ComputeTarget
from azureml.core.environment import Environment
from azureml.train.estimator import Estimator
from IPython.display import Image, display
from sklearn.preprocessing import MinMaxScaler
from statsmodels.tsa.statespace.sarimax import SARIMAX

get_ipython().run_line_magic("matplotlib", "inline")
pd.options.display.float_format = "{:,.2f}".format
np.set_printoptions(precision=2)
warnings.filterwarnings("ignore")  # specify to ignore warning messages
```

For the second step, you need to configure your workspace. You can set up your Azure Machine Learning (Azure ML) service (aka.ms/AzureMLservice) workspace and configure your notebook library by running the following code:

```
# Configure the workspace, if no config file has been downloaded.
subscription_id = os.getenv("SUBSCRIPTION_ID", default="<Your Subcription ID>")
resource_group = os.getenv("RESOURCE_GROUP", default="<Your Resource Group>")
workspace_name = os.getenv("WORKSPACE_NAME", default="<Your Workspace Name>")
workspace_region = os.getenv
("WORKSPACE_REGION", default="<Your Workspace Region>")

try:
    ws = Workspace(subscription_id = subscription_id,
                    resource_group = resource_group,
                    workspace_name = workspace_name)
    ws.write_config()
    print("Workspace configuration succeeded")
```

```
except:
print("Workspace not accessible.
Change your parameters or create a new workspace below")


# Or take the configuration of the existing config.json file
ws = Workspace.from_config()
print(ws.name, ws.resource_group, ws.location, ws.subscription_id, sep='\n')
```

Make sure that you have the correct version of Azure ML SDK. If that's not the case, you can run the following code:

```
!pip install --upgrade azureml-sdk[automl,notebooks,explain]
!pip install --upgrade azuremlftk
```

Then configure your workspace and write the configuration to a config.json file or read your config.json file to get your workspace. As a second option, you can copy the config file from the Azure workspace in an azureml folder.

In an Azure workspace you will find the following items:

- Experiment results
- Trained models
- Compute targets
- Deployment containers
- Snapshots

For more information about the AML services workspace setup, visit aka.ms/ AzureMLConfiguration and follow the instructions in the notebook.

Create an Experiment

We now create an Azure Machine Learning experiment, which will help keep track of the specific data used as well as the model training job logs. If the experiment already exists on the selected workspace, the run will be added to the existing experiment. If not, the experiment will be added to the workspace, as shown in the following code:

```
experiment_name = 'energydemandforecasting'
exp = Experiment(workspace=ws, name=experiment_name)
```

Create or Attach a Compute Cluster

At this point, you need to create or attach an existing compute cluster. For training an ARIMA model, a CPU cluster is enough, as illustrated in the code below. Note the `min_nodes` parameter is 0, meaning by default this will have no machines in the cluster:

```
# choose a name for your cluster
compute_name = os.environ.get("AML_COMPUTE_CLUSTER_NAME", "cpucluster")

compute_min_nodes = os.environ.get("AML_COMPUTE_CLUSTER_MIN_NODES", 0)
compute_max_nodes = os.environ.get("AML_COMPUTE_CLUSTER_MAX_NODES", 4)

# This example uses CPU VM. For using GPU VM, set SKU to STANDARD_NC6
vm_size = os.environ.get("AML_COMPUTE_CLUSTER_SKU", "STANDARD_D2_V2")

if compute_name in ws.compute_targets:
    compute_target = ws.compute_targets[compute_name]
    if compute_target and type(compute_target) is AmlCompute:
        print('found compute target. just use it. ' + compute_name)
else:
    print('creating a new compute target...')
    provisioning_config =
        AmlCompute.provisioning_configuration
        (vm_size = vm_size,
        min_nodes = compute_min_nodes,
        max_nodes = compute_max_nodes)

    # create the cluster
    compute_target = ComputeTarget.create(ws,
    compute_name,  provisioning_config)

    # can poll for a minimum number of nodes and for a specific timeout.
    compute_target.wait_for_completion
    (show_output=True, min_node_count=None,
    timeout_in_minutes=20)

    # For a more detailed view of current
    # AmlCompute status, use 'get_status()'
    print(compute_target.get_status().serialize())
```

Upload the Data to Azure

Now you need to make the data accessible remotely by uploading it from your local machine into Azure. Then it can be accessed for remote training. The datastore is a convenient construct associated with your workspace for you to upload or download data. You can also interact with it from your remote compute targets. It's backed by an Azure Blob storage account. The energy file is uploaded into a directory named `energy_data` at the root of the datastore:

▪ First, you can download the GEFCom2014 data set and save the files into a data directory locally, which can be done by executing the commented lines in the cell. The data in this example is taken from the GEFCom2014 forecasting competition. It consists of three years of hourly electricity load and temperature values between 2012 and 2014.

▪ Then, the data is uploaded to the default blob data storage attached to your workspace. The energy file is uploaded into a directory named `energy_data` at the root of the datastore. The upload of data must be run only the first time. If you run it again, it will skip the uploading of files already present on the datastore.

```
# save the files into a data directory locally
data_folder = './data'

#data_folder = os.path.join(os.getcwd(), 'data')
os.makedirs(data_folder, exist_ok=True)

# get the default datastore
ds = ws.get_default_datastore()
print(ds.name, ds.datastore_type, ds.account_name, ds.container_name,
sep='\n')

# upload the data
ds.upload(src_dir=data_folder,
target_path='energy_data',
overwrite=True,
show_progress=True)

ds = ws.get_default_datastore()
print(ds.datastore_type, ds.account_name, ds.container_name)
```

Now we need to create a training script:

```
# ## Training script
# This script will be given to the estimator
# which is configured in the AML training script.
# It is parameterized for training on `energy.csv` data.

#%% [markdown]
# ### Import packages.
# utils.py needs to be in the same directory as this script,
# i.e., in the source directory `energydemandforcasting`.

#%%
import argparse
```

Continues

(continued)

```
import os
import numpy as np
import pandas as pd
import azureml.data
import pickle

from statsmodels.tsa.statespace.sarimax import SARIMAX
from sklearn.preprocessing import MinMaxScaler
from utils import load_data, mape
from azureml.core import Run

#%% [markdown]
# ### Parameters
# * COLUMN_OF_INTEREST: The column containing data that will be
forecasted.
# * NUMBER_OF_TRAINING_SAMPLES:
#    The number of training samples that will be trained on.
# * ORDER:
#    A tuple of three non-negative integers
#    specifying the parameters p, d, q of an Arima(p,d,q) model,
#    where:
#        * p: number of time lags in autoregressive model,
#        * d: the degree of differencing,
#        * q: order of the moving avarage model.
# * SEASONAL_ORDER:
#    A tuple of four non-negative integers
#    where the first three numbers
#    specify P, D, Q of the Arima terms
#    of the seasonal component, as in ARIMA(p,d,q)(P,D,Q).
#    The fourth integer specifies m,
#    i.e, the number of periods in each season.

#%%
COLUMN_OF_INTEREST = 'load'
NUMBER_OF_TRAINING_SAMPLES = 2500
ORDER = (4, 1, 0)
SEASONAL_ORDER = (1, 1, 0, 24)

#%% [markdown]
# ### Import script arguments
# Here, Azure will read in the parameters, specified in the AML
training.

#%%
parser = argparse.ArgumentParser(description='Process input arguments')
parser.add_argument
('--data-folder',
default='./data/',
type=str,
```

```python
            dest='data_folder')
parser.add_argument
('--filename',
default='energy.csv',
type=str,
dest='filename')
parser.add_argument('--output', default='outputs', type=str, dest='output')
args = parser.parse_args()
data_folder = args.data_folder
filename = args.filename
output = args.output
print('output', output)
#%% [markdown]
# ### Prepare data for training
# * Import data as pandas dataframe
# * Set index to datetime
# * Specify the part of the data that the model will be fitted on
# * Scale the data to the interval [0, 1]

#%%
# Import data
energy = load_data(os.path.join(data_folder, filename))
# As we are dealing with time series, the index can be set to datetime.
energy.index = pd.to_datetime(energy.index, infer_datetime_format=True)

# Specify the part of the data that the model will be fitted on.
train = energy.iloc[0:NUMBER_OF_TRAINING_SAMPLES, :]

# Scale the data to the interval [0, 1].
scaler = MinMaxScaler()
train[COLUMN_OF_INTEREST] =
scaler.fit_transform(np.array
(train.loc[:, COLUMN_OF_INTEREST].values).
reshape(-1, 1))
#%% [markdown]
# ### Fit the model

#%%
model = SARIMAX(endog=train[COLUMN_OF_INTEREST]
.tolist(),
order=ORDER,
seasonal_order=SEASONAL_ORDER)
model.fit()

#%% [markdown]
# ### Save the model
# The model will be saved on Azure in the specified directory as a pickle file.

#%%
# Create a directory on Azure in which the model will be saved.
os.makedirs(output, exist_ok=True)
```

Continues

(continued)
```
# Write the the model as a .pkl file to the specified directory on Azure.
with open(output + '/arimamodel.pkl', 'wb') as m:
    pickle.dump(model, m)

# with open('arimamodel.pkl', 'wb') as m:
#     pickle.dump(model, m)

#%%
```

Create an Estimator

Let's see now how to create an estimator. In order to start this process, we need to create some parameters. The following parameters will be given to the estimator:

- `source_directory`: the directory that will be uploaded to Azure and contains the script `train.py`
- `entry_script`: the script that will be executed (`train.py`)
- `script_params`: the parameters that will be given to the entry script
- `compute_target`: the compute cluster that was created above
- `conda_dependencies_file`: the packages in your `conda` environment that the script needs.

```
script_params = {
    "--data-folder": ds.path("energy_data").as_mount(),
    "--filename": "energy.csv",
}
script_folder = os.path.join(os.getcwd(), "energydemandforecasting")

est = Estimator(
    source_directory=script_folder,
    script_params=script_params,
    compute_target=compute_target,
    entry_script="train.py",
    conda_dependencies_file="azureml-env.yml",
)
```

Submit the Job to the Remote Cluster

You can create and manage a compute target using the Azure Machine Learning SDK, Azure portal, Azure CLI, or Azure Machine Learning VS Code extension. The following sample code shows you how to submit your work to the remote cluster:

```
run = exp.submit(config=est)

# specify show_output to True for a verbose log
run.wait_for_completion(show_output=False)
```

Register the Model

The last step in the training script wrote the file `outputs/arima_model.pkl` in a directory named outputs in the VM of the cluster where the job is run. Outputs is a special directory in that all content in this directory is automatically uploaded to your workspace. This content appears in the run record in the experiment under your workspace. So the model file is now also available in your workspace.

You can also see files associated with that run. As a last step, we register the model in the workspace, which saves it under Models on Azure, so that you and other collaborators can later query, examine, and deploy this model. By registering the model, it is now available on your workspace:

```
# see files associated with that run
print(run.get_file_names())

# register model
model = run.register_model(model_name='arimamodel',
model_path='outputs/arimamodel.pkl')
```

Deployment

Once we have nailed down the modeling phase and validated the model performance, we are ready to go into the deployment phase. In this context, *deployment* means enabling the customer to consume the model by running actual predictions on it at large scale. The concept of deployment is key in Azure ML since our main goal is to constantly invoke predictions as opposed to just obtaining the insight from the data. The deployment phase is the part where we enable the model to be consumed at large scale.

Within the context of energy demand forecast, our aim is to invoke continuous and periodical forecasts while ensuring that fresh data is available for the model and that the forecasted data is sent back to the consuming client.

The main deployable building block in Azure ML is the web service. This is the most effective way to enable consumption of a predictive model in the cloud. The web service encapsulates the model and wraps it up with a REST API (application programming interface). The API can be used as part of any client code, as illustrated in Figure 6.6.

The web service is deployed on the cloud and can be invoked over its exposed REST API endpoint, which you can see in Figure 6.6. Different types of clients across various domains can invoke the service through the Web API simultaneously. The web service can also scale to support thousands of concurrent calls.

Deploying the model requires the following components:

▪ An entry script. This script accepts requests, scores the request using the model, and returns the results.

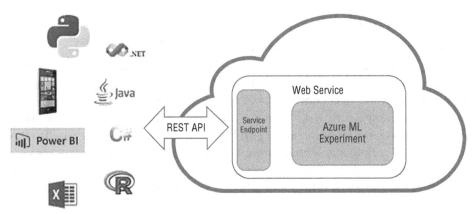

Figure 6.6: Web service deployment and consumption

- Dependencies, such as helper scripts or Python/Conda packages required to run the entry script or model.
- The deployment configuration for the compute target that hosts the deployed model. This configuration describes things like memory and CPU requirements needed to run the model.

These entities are encapsulated into an inference configuration and a deployment configuration. The inference configuration references the entry script and other dependencies. These configurations are defined programmatically when using the SDK and as JSON files when using the CLI to perform the deployment.

Define Your Entry Script and Dependencies

The entry script receives data submitted to a deployed web service and passes it to the model. It then takes the response returned by the model and returns that to the client. The script is specific to your model; it must understand the data that the model expects and returns.

The script contains two functions that load and run the model: the `init()` and `run(input_data)` functions. When you register a model, you provide a model name used for managing the model in the registry. You use this name with the `Model.get_model_path()` to retrieve the path of the model file(s) on the local file system. If you register a folder or a collection of files, this API returns the path to the directory that contains those files.

When you register a model, you give it a name which corresponds to where the model is placed, either locally or during service deployment. The following example will return a path to a single file called `sklearn_mnist_model.pkl` (which was registered with the name `sklearn_mnist`):

```
model_path = Model.get_model_path('sklearn_mnist')
```

Automatic Schema Generation

To automatically generate a schema for your web service, you need to provide a sample of the input and output in the constructor for one of the defined type objects, and the type and sample are used to automatically create the schema (aka.ms/ModelDeployment).

To use schema generation, include the inference-schema package in your Conda environment file. After this, you need to define the input and output sample formats in the `input_sample` and `output_sample` variables. The following example demonstrates how to accept and return JSON data for our energy demand forecasting solution. First, the workspace that was used for training must be retrieved:

```
ws = Workspace.from_config()
print(ws.name, ws.resource_group, ws.location, ws.subscription_id, sep = '\n')
```

We already registered the model in the training script. But if the model you want to use is only saved locally, you can uncomment and run the following cell that will register your model in the workspace. Parameters may need adjustment:

```
# model = Model.register(model_path = "path_of_your_model",
#                        model_name = "name_of_your_model",
#                        tags = {'type': "Time series ARIMA model"},
#                        description = "Time series ARIMA model",
#                        workspace = ws)

# get the already registered model
model = Model.list(ws, name='arimamodel')[0]
print(model)
```

We now need to get or register an environment for our model deployment (aka.ms/ModelDeployment). Since, in our example, we already registered the environment in the training script, we can just retrieve it:

```
my_azureml_env = Environment.get(workspace=ws, name="my_azureml_env")

inference_config = InferenceConfig(
    entry_script="energydemandforecasting/score.py", environment=my_
azureml_env
)
```

After this, the deployment configuration can be arranged, as illustrated in the sample code below:

```
# Set deployment configuration
deployment_config = AciWebservice.deploy_configuration(cpu_cores=1, memory_gb=1)

aci_service_name = "aci-service-arima"
```

Finally, the deployment configuration and web service name and location to deploy can be defined, as illustrated in the sample code below:

```
# Define the web service
service = Model.deploy(
    workspace=ws,
    name=aci_service_name,
    models=[model],
    inference_config=inference_config,
    deployment_config=deployment_config,
)
service.wait_for_deployment(True)
```

Below is an overview of the code in the scoring file, named `score.py`:

```
### score.py
#### Import packages
import pickle
import json
import pandas as pd
from sklearn.preprocessing import MinMaxScaler

from azureml.core.model import Model

MODEL_NAME = 'arimamodel'
DATA_NAME = 'energy'
DATA_COLUMN_NAME = 'load'
NUMBER_OF_TRAINING_SAMPLES = 2500
HORIZON = 10

#### Init function
def init():
    global model
    model_path = Model.get_model_path(MODEL_NAME)
    # deserialize the model file back into a sklearn model
    with open(model_path, 'rb') as m:
        model = pickle.load(m)

#### Run function
def run(energy):
    try:
        # load data as pandas dataframe from the json object.
        energy = pd.DataFrame(json.loads(energy)[DATA_NAME])
        # take the training samples
        energy = energy.iloc[0:NUMBER_OF_TRAINING_SAMPLES, :]

        scaler = MinMaxScaler()
        energy[DATA_COLUMN_NAME] =
        scaler.fit_transform
```

```
        (energy[[DATA_COLUMN_NAME]])
        model_fit = model.fit()

        prediction = model_fit.forecast(steps = HORIZON)
        prediction = pd.Series.to_json
        (pd.DataFrame(prediction),
        date_format='iso')

        # you can return any data type as long as it is JSON-serializable
        return prediction
    except Exception as e:
        error = str(e)
        return error
```

Before deploying, you must define the deployment configuration. The deployment configuration is specific to the compute target that will host the web service. The deployment configuration is not part of your entry script. It is used to define the characteristics of the compute target that will host the model and entry script (aka.ms/ModelDeployment). You may also need to create the compute resource—for example, if you do not already have an Azure Kubernetes Service associated with your workspace.

Table 6.1 provides an example of creating a deployment configuration for each compute target:

Table 6.1: Creating a deployment configuration for each compute target

COMPUTE TARGET	DEPLOYMENT CONFIGURATION EXAMPLE
Number of data points	`deployment_config = LocalWebservice.` `deploy_configuration(port=8890)`
Azure Container Instances	`deployment_config = AciWebservice.deploy_` `configuration(cpu_cores = 1, memory_gb = 1)`
Azure Kubernetes Service	`deployment_config = AksWebservice.deploy_` `configuration(cpu_cores = 1, memory_gb = 1)`

For this specific example, we are going to create an Azure Container Instances (ACI), which is typically used when you need to quickly deploy and validate your model or you are testing a model that is under development.

First, you must configure the service with the number of CPU cores, the size of the memory, and other parameters like the description. Then, you must deploy the service from the image.

You can deploy the service only once. If you want to deploy it again, change the name of the service or delete the existing service directly on Azure:

```
# load the data to use for testing and encode it in json
energy_pd = load_data('./data/energy.csv')
energy = pd.DataFrame.to_json(energy_pd, date_format='iso')
```

Continues

(continued)

```
energy = json.loads(energy)
energy = json.dumps({"energy":energy})

# Call the service to get the prediction for this time series
prediction = aci_service.run(energy)
```

If you want, at this final step, you can plot the prediction results. The following sample will help you achieve the following three tasks:

- Convert the prediction to a DataFrame containing correct indices and columns.
- Scale the original data as in the training.
- Plot the original data and the prediction.

```
# prediction is a string, convert it to a dictionary
prediction = ast.literal_eval(prediction)

# convert the dictionary to pandas dataframe
prediction_df = pd.DataFrame.from_dict(prediction)

prediction_df.columns=['load']
prediction_df.index = energy_pd.iloc[2500:2510].index)

# Scale the original data
scaler = MinMaxScaler()
energy_pd['load'] = scaler.fit_transform
(np.array(energy_pd.loc[:, 'load'].values).
reshape(-1, 1))

# Visualize a part of the data before the forecasting
original_data = energy_pd.iloc[1500:2501]

# Plot the forecasted data points
fig = plt.figure(figsize=(15, 8))

plt.plot_date(x=original_data.index,
y=original_data, fmt='-',
xdate=True, label="original load", color='red')
plt.plot_date(x=prediction_df.index, y=prediction_df, fmt='-',
xdate=True, label="predicted load", color='yellow')
```

When deploying an energy demand forecasting solution, we are interested in deploying an end-to-end solution that goes beyond the prediction web service and facilitates the entire data flow. At the time we invoke a new forecast, we would need to make sure that the model is fed with the up-to-date data features. That implies that the newly collected raw data is constantly ingested, processed, and transformed into the required feature set on which the model was built.

At the same time, we would like to make the forecasted data available for the end consuming clients. An example data flow cycle (or data pipeline) is illustrated in Figure 6.7.

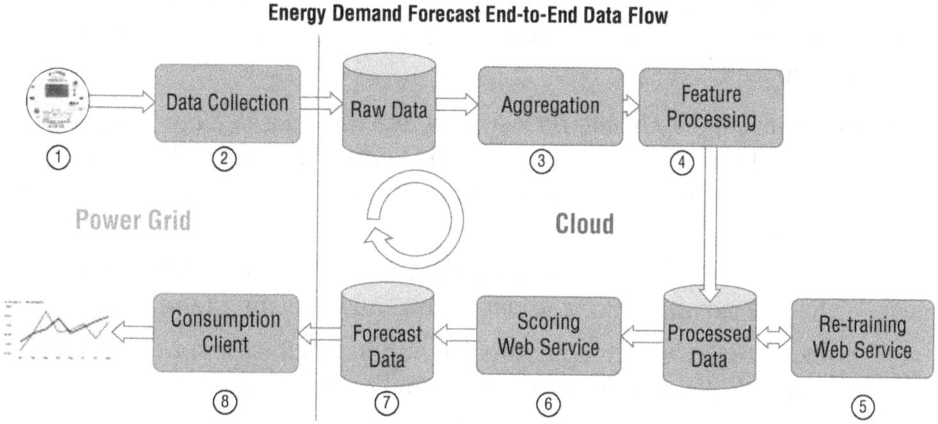

Energy Demand Forecast End-to-End Data Flow

Figure 6.7: Energy demand forecast end-to-end data flow

These are the steps that take place as part of the energy demand forecast cycle:

1. Millions of deployed data meters are constantly generating power consumption data in real time.

2. This data is being collected and uploaded onto a cloud repository (such as Azure Storage).

3. Before being processed, the raw data is aggregated to a substation or regional level as defined by the business.

4. The feature processing then takes place and produces the data that is required for model training or scoring—the feature set data is stored in a database (such as Azure SQL Database).

5. The retraining service is invoked to retrain the forecasting model. That updated version of the model is persisted so that it can be used by the scoring web service.

6. The scoring web service is invoked on a schedule that fits the required forecast frequency.

7. The forecasted data is stored in a database that can be accessed by the end consumption client.

8. The consumption client retrieves the forecasts and applies it back into the grid and consumes it in accordance with the required use case.

It is important to note that this entire cycle is fully automated and runs on a schedule.

Conclusion

In this final chapter, we looked closely into the process of building and deploying some of the time series forecasting solutions. Specifically, this chapter provided a complete overview of tools to build and deploy your own time series forecasting solutions:

- *Experimental Set Up and Introduction to Azure Machine Learning SDK for Python* – I introduced Azure Machine Learning SDK for Python to build and run machine learning workflows, and you learned the following concepts and assets:

  - `Workspace`, which is foundational resource in the cloud that you use to experiment, train, and deploy machine learning models.

  - `Experiment`, which is another foundational cloud resource that represents a collection of trials (individual model runs).

  - `Run`, which represents a single trial of an experiment.

  - `Model`, which is used for working with cloud representations of the machine learning model.

  - `ComputeTarget, RunConfiguration, ScriptRunConfig`, which are abstract parent classes for creating and managing compute targets. A compute target represents a variety of resources where you can train your machine learning models.

  - `Image`, which is abstract parent class for packaging models into container images that include the runtime environment and dependencies.

  - `Webservice`, which is the abstract parent class for creating and deploying web services for your models.

- *Machine Learning Model Deployment* – This section introduced the machine learning model deployment, that is, the method of integrating a machine learning model into an existing production environment in order to begin developing practical business decisions based on data.

- *Solution Architecture for Time Series Forecasting with Deployment Examples* – In this final section, we built, trained, and deployed an end-to-end data pipeline architecture and walked through the deployment code and examples.

References

Bianchi, Filippo Maria, and Enrico Maiorino, Michael Kampffmeyer, Antonello Rizzi, Robert Jenssen. 2018. *Recurrent Neural Networks for Short-Term Load Forecasting*. Berlin, Germany: Springer.

Brownlee, Jason. 2017. *Introduction to Time Series Forecasting With Python - Discover How to Prepare Data and Develop Models to Predict the Future*. Machine Learning Mastery. https://machinelearningmastery.com/introduction-to-time-series-forecasting-with-python/.

Che, Zhengping, and Sanjay Purushotham, Kyunghyun Cho, David Sontag, Yan Liu. 2018. "Recurrent Neural Networks for Multivariate Time Series with Missing Values." *Scientific Reports 8*. https://doi.org/10.1038/s41598-018-24271-9.

Cheng H., Tan PN., Gao J., Scripps J. 2006. "Multistep-Ahead Time Series Prediction." In: Ng WK., Kitsuregawa M., Li J., Chang K. (eds) *Advances in Knowledge Discovery and Data Mining*, PAKDD 2006. Lecture Notes in Computer Science 3918. *Berlin, Heidelberg: Springer*. https://doi.org/10.1007/11731139_89.

Cho, Kyunghyun, and Bart van Merriënboer, Caglar Gulcehre, Fethi Bougares, Holger Schwenk, Y Bengio. 2014. *Learning Phrase Representations using RNN Encoder-Decoder for Statistical Machine Translation*. https://doi.org/10.3115/v1/D14-1179.

Glen, Stephanie. 2014. "Endogenous Variable and Exogenous Variable: Definition and Classifying." Statistics How To blog. https://www.statisticshowto.com/endogenous-variable/.

Lazzeri, Francesca. 2019a. "3 reasons to add deep learning to your time series toolkit."O'Reilly Ideas blog. https://www.oreilly.com/content/3-reasons-to-add-deep-learning-to-your-time-series-toolkit/.

Lazzeri, Francesca. 2019b. "Data Science Mindset: Six Principles to Build Healthy Data-Driven Organizations." InfoQ blog. `https://www.infoq .com/articles/data-science-organization-framework`.

Lazzeri, Francesca. 2019c. "How to deploy machine learning models with Azure Machine Learning." Educative.io blog. `https://www.educative. io/blog/how-to-deploy-your-machine-learning-model`.

Lewis-Beck, Michael S., and Alan Bryman, Tim Futing Liao. 2004. *The Sage Encyclopedia of Social Science Research Methods*. Thousand Oaks, Calif: Sage.

Nguyen, Giang, and Stefan Dlugolinsky, Martin Bobak, Viet Tran, Alvaro Lopez Garcia, Ignacio Heredia, Peter Malík, Ladislav Hluchý. 2019. "Machine Learning and Deep Learning frameworks and libraries for large-scale data mining: a survey". *Artificial Intelligence Review* 52: 77–124. https://doi.org/ 10.1007/s10462-018-09679-z.

Petris, G., and S. Petrone, P. Campagnoli. 2009. *Dynamic Linear Models in R*. Springer.

Poznyak, T., and J.I.C. Oria, A. Poznyak. 2018. *Ozonation and Biodegradation in Environmental Engineering: Dynamic Neural Network Approach*. Elsevier Science.

Stellwagen, Eric. 2011. *"Forecasting 101: A Guide to Forecast Error Measurement Statistics and How to Use Them."* ForecastPRO blog. `https://www .forecastpro.com/Trends/forecasting101August2011.html`.

Hong, Tao, and Pierre Pinson, Fan Shu, Hamidreza Zareipour, Alberto Troccoli, Rob J. Hyndman. 2016. "Probabilistic Energy Forecasting: Global Energy Forecasting Competition 2014 and Beyond." *International Journal of Forecasting* 32, no. 3 (July-September): 896–913.

Taylor, Christine. 2018. "Structured vs. Unstructured Data." Datamation blog. `https://www.datamation.com/big-data/structured-vs-unstructured- data.html`.

White, Halbert. 1980. "A Heteroskedasticity-Consistent Covariance Matrix Estimator and a Direct Test for Heteroskedasticity." *Econometrica, Econometric Society* 48, no.4 (May):817–838.

Zhang, Ruiyang, and Zhao Chen, Chen Su, Jingwei Zheng, Oral Büyüköztürk, Hao Sun. 2019. "Deep long short-term memory networks for nonlinear structural seismic response prediction." *Computers & Structures* 220: 55–68. https://doi.org/10.1016/j.compstruc.2019.05.006.

Zhang Y, and YT Zhang, JY Wang, XW Zheng. 2015. "Comparison of classification methods on EEG signals based on wavelet packet decomposition." *Neural Comput Appl 26*, no. 5:1217–1225.

Zuo, Jingwei, and Karine Zeitouni, Yehia Taher. 2019. ISETS: Incremental Shapelet Extraction from Time Series Stream. In: Brefeld U., Fromont E., Hotho A., Knobbe A., Maathuis M., Robardet C. (eds) Machine Learning and Knowledge Discovery in Databases. ECML PKDD 2019. *Lecture Notes in Computer Science* 11908: Springer, Cham. https://doi.org/ 10.1007/978-3-030-46133-1_53.

Index